Petra Thoma

Ultra-fast YBa$_2$Cu$_3$O$_{7-x}$ direct detectors for the THz frequency range

Ultra-fast YBa$_2$Cu$_3$O$_{7-x}$ direct detectors for the THz frequency range

BAND 009

Karlsruher Schriftenreihe zur Supraleitung

Herausgeber

Prof. Dr.-Ing. M. Noe
Prof. Dr. rer. nat. M. Siegel

Eine Übersicht über alle bisher in dieser Schriftenreihe
erschienene Bände finden Sie am Ende des Buchs.

Ultra-fast YBa$_2$Cu$_3$O$_{7-x}$ direct detectors for the THz frequency range

by
Petra Thoma

Dissertation, Karlsruher Institut für Technologie (KIT)
Fakultät für Elektrotechnik und Informationstechnik, 2013
Hauptreferent: Prof. Dr. Michael Siegel
Korreferent: Prof. Dr. Shaukat Khan

Impressum

Karlsruher Institut für Technologie (KIT)
KIT Scientific Publishing
Straße am Forum 2
D-76131 Karlsruhe
www.ksp.kit.edu

KIT – Universität des Landes Baden-Württemberg und
nationales Forschungszentrum in der Helmholtz-Gemeinschaft

KIT Scientific Publishing 2013
Print on Demand

ISSN 1869-1765
ISBN 978-3-7315-0070-4

Ultra-fast $YBa_2Cu_3O_{7-x}$ direct detectors for the THz frequency range

Zur Erlangung des akademischen Grades eines

DOKTOR-INGENIEURS

von der Fakultät für
Elektrotechnik und Informationstechnik
des Karlsruher Instituts für Technologie (KIT)
genehmigte

DISSERTATION

von

Dipl.-Ing. Petra Thoma, geb. Probst

geboren in Sigmaringen

Tag der mündlichen Prüfung: 21.Juni 2013
Hauptreferent: Prof. Dr. Michael Siegel
Korreferent: Prof. Dr. Shaukat Khan

Preface

This dissertation is the result of my work at the Institut für Mikro- und Nanoelektronische Systeme (IMS) at the Karlsruhe Institute of Technology (KIT). This thesis would have not been possible without the help of a number of people, who I want to thank in the following. After that, details about the way citations are given within this thesis are discussed.

Danksagung

Als erstes möchte ich mich herzlich bei Prof. Siegel für die Möglichkeit am Institut für Mikro- und Nanoelektronische Systeme zu promovieren, bedanken. Er hat mich stets auf meinem Weg gefördert und gefordert und mich bei meinen Vorhaben in Zusammenhang mit der Promotionsarbeit unterstützt. Das mir entgegengebrachte Vertrauen, aber auch die geäußerten Erwartungen haben mich auf dem Weg der Promotion gestärkt.

Des Weiteren möchte ich mich recht herzlich bei Prof. Khan für die Übernahme des Korreferats bedanken, wodurch die Arbeit auch hinsichtlich der Anwendung detailliert beleuchtet wurde, was mich sehr gefreut hat.

Von ganzem Herzen möchte ich meinem Betreuer Dr. Konstantin Il'in danken. Er hat durch seine jahrelange, unermüdliche Unterstützung dazu beigetragen, dass ich auch in schwierigen Phasen der Arbeit einen Ausweg gefunden habe. Er hatte stets ein offenes Ohr für meine Fragen und ließ mich an seiner umfassenden Erfahrung teilhaben.

Einen ganz lieben Dank möchte ich an meine Kollegen richten. Hierbei möchte ich besonders Alexander Scheuring danken, der immer dafür gesorgt hat, dass die Arbeit am IMS und insbesondere die Messkampagnen bei ANKA kurzweilig und erfolgreich waren. Ebenso möchte ich Dr. Stefan Wünsch danken, der mir immer ein guter elektrotechnischer Ratgeber war und mit viel Humor und Unterstützung meine Promotionszeit betreut hat. Max Meckbach danke ich für die arbeitsintensiven und trotzdem kurzweiligen Arbeitsstunden in der Technologie und freue mich jetzt schon, wenn wir uns wieder auf Proisland sehen. Für die unermüdliche Unterstützung im Bereich angewandter Elektrotechnik möchte ich Matthias Hofherr und Matthias Arndt danken, die mir viele unterhaltsame Stunden in der Elektronikwerkstatt beschert haben. Außerdem möchte ich Erich Crocoll, Doris Duffner, Dr. Gerd Hammer, Dagmar Henrich, Dr. Christoph Kaiser, Michael Merker, Axel Stockhausen und Philipp Trojan für die tolle Arbeitsatmosphäre am IMS danken! Besonders bedanken möchte ich mich bei der technischen Unterstützung am Institut. Herr Gutbrod, Herr Stassen und Herr Wermund waren mir eine unverzichtbare Hilfe während der Promotion. Bei Herrn Gutbrod möchte ich mich ganz herzlich für seine Geduld bei meinen technischen Zeichnungen bedanken als auch seinem Erfindungsreichtum und Mut, sich immer wieder neuen

Ideen von mir zu stellen. Herr Stassen hat mir durch sein Feingefühl und Geschick beim Bonden der Detektoren viele graue Haare erspart und zum erfolgreichen Gelingen der Messungen beigetragen. Herr Wermund hat meine ersten Schritte am Laser begleitet und durch seine Frohnatur dazu beigetragen, dass ich in verbissenen Momenten wieder etwas entspannter in der Technologie stand. Ebenso möchte ich bei den zahlreichen Studenten bedanken, die durch ihre tatkräftige Mitarbeit diese Arbeit unterstützt haben. Hierbei danke ich besonders meiner jetzigen Kollegin Juliane Raasch, die durch ihre Neugierde und Engagement auch mich immer wieder angestachelt hat Unbekanntes zu erforschen und Bestehendes zu hinterfragen. Ebenso danken möchte ich Julia Day, Diana Kalteisen, Mikel de Zabala Wissing und Holger Wernerus für die enge Zusammenarbeit.

Es ist mir ein besonderes Anliegen Prof. Alexei Semenov für seine Unterstützung während meiner Promotion zu danken. Ohne seinen Beitrag wäre die Arbeit in dieser Form nicht zustande gekommen. Des Weiteren möchte ich Prof. Heinz-Wilhelm Hübers sowie dem gesamten Team an der Metrology Light Source in Berlin für Ihre Unterstützung danken.

An das Team von ANKA geht mein Dank für die tatkräftige Unterstützung während den letzten drei Jahren bei unseren zahlreichen Messungen. Prof. Anke-Susanne Müller danke ich für die Unterstützungen während den Messkampagnen bei ANKA, welche entscheidend zum erfolgreichen Gelingen der Arbeit beigetragen hat. Vitali Judin möchte ich für die Hilfe bei der anschließenden Datenanalyse, sowie die zahlreichen Diskussionen danken. Ebenso haben Nicole Hiller, Michael Süpfle und Johannes Steinmann zum erfolgreichen Gelingen der Arbeit an ANKA entscheidend beigetragen. Dr. Nigel Smale und Dr. Erhard Huttel möchte ich für den unermüdlichen Einsatz im Kontrollraum danken, der weit über das Selbstverständliche hinausging.

Auch dem Team von UVSOR in Japan möchte ich für die erfolgreichen Messungen der letzten Jahre danken. Prof. Katoh und Prof. Kimura haben durch ihr Interesse an einer Zusammenarbeit unserer Gruppen spannende Messkampagnen an UVSOR ermöglicht. Hosaka-san, Zen-san, Konomi-san sowie Prof. Serge Bielawski, Dr. Clément Evain, Dr. Christophe Szwaj und Eléonore Roussel möchte ich für die angenehme und produktive Kollaboration danken.

Ebenso möchte ich mich für die anhaltende Unterstützung mit den schnellsten Oszilloskopen bei Markus Stocklas und Matthias Kohler von Agilent Technologies bedanken, welche einige Messungen dieser Arbeit erst möglich machten. Ebenso danke ich Prof. Zwick und Jochen Antes vom IHE, KIT für die wiederholte Leihgabe des großartigen Oszilloskops.

Zuletzt möchte ich mich von ganzem Herzen bei meiner Familie und Freunden bedanken, die während den letzten drei Jahren stets Verständnis zeigten, wenn ich aufgrund von wichtigen Reisen und Messkampagnen nicht für sie da sein konnte. Mein größter Dank geht an meinen Mann Michael, der mich stets bedingungslos unterstützt hat. Sein unerschütterlicher Glaube an den Erfolg meiner Arbeit haben entscheidend zum Gelingen dieser Promotion beigetragen. Danke, dass du an meiner Seite bist!

Karlsruhe,   *Petra Thoma*
im April 2013   *Karlsruher Institut für Technologie (KIT)*

Citations

The references cited in this thesis are separated in four classes and are represented by different abbreviations:

External publications are given by numbers in square brackets, sorted by their first occurrence in the text. Example:

[1] A.-S. Müller. Accelerator-Based Sources of Infrared and Terahertz Radiation. *Reviews of Accelerator Science and Technology*, 3(1):165-183, 2010.

My own publications in scientific journals are given in a separate list. They are cited as the first letters of the last names of the first 3 authors, followed by the year of publication, all in square brackets. Example:

[TSH+12] P. Thoma, A. Scheuring, M. Hofherr, S. Wünsch, K. Il'in, N. Smale, V. Judin, N. Hiller, A.-S. Müller, A. Semenov, H.-W. Hübers, and M. Siegel. Real-time measurements of picosecond THz pulses by an ultra-fast $YBa_2Cu_3O_{7-d}$ detection system. *Applied Physics Letters*, 101 (142601), 2012.

A list of the international conferences I have attended during my thesis is the third category. They are referred to by a letter, giving the chronological order, followed by the name of the conference. Example:

[a-ASC] P. Thoma, J. Raasch, A. Scheuring, M. Hofherr, K. Il'in, S. Wünsch, A. Semenov, H.-W. Hübers, V. Judin, A.-S. Müller, N. Smale, J. Hänisch, B. Holzapfel, and M. Siegel. Thin YBCO film THz detector with picosecond time resolution and large dynamic range. *Invited talk at the Applied Superconductivity Conference 2012 (ASC)*, Portland, Oregon, US, 7-12 of October 2012.

Student theses, which I supervised during my PhD thesis, make the fourth list. They are cited as the first three letters of the last name of the student, followed by the year of completion. Example:

[Raa12] Raasch, Juliane. *Detektionsmechanismus in sub-Mikrometer YBCO Strukturen*, 2012. Diploma thesis, Institut für Mikro- und Nanoelektronische Systeme, Karlsruher Institut für Technologie (KIT).

The complete lists are given at the end of the thesis.

Zusammenfassung

Diese Arbeit beschreibt die Entwicklung von Direktdetektoren für den Terahertz-Frequenzbereich basierend auf Dünnschichten des Hochtemperatursupraleiters $YBa_2Cu_3O_{7-x}$ (YBCO) mit einer zeitlichen Auflösung im Pikosekundenbereich und deren Einbettung in ein neu entwickeltes ultraschnelles Auslesesystem, welches in der Lage ist, zeitliche Prozesse innerhalb dieser Zeitskalen aufzulösen.

Es gibt bereits eine Reihe von Terahertz-Direktdetektoren aus Metallen, Halbleitern und Supraleitern, wobei gekühlte Detektoren stets den Vorteil der höheren Sensitivität aufweisen. Neben der Sensitivität ist die Antwortzeit eines Detektors ein wichtiges Charakteristikum, welches das dynamische Verhalten des Detektors beschreibt. Während die meisten Direktdetektoren Antwortzeiten im Mikro- und Millisekundenbereich aufweisen, erlauben lediglich zwei Detektortechnologien die Auflösung ultraschneller Prozesse im Pikosekundenbereich. Dies sind zum einen Schottkydioden, welche intrinsische Antwortzeiten von wenigen Pikosekunden aufweisen, als auch supraleitende Hot-Electron-Bolometer, deren Antwortzeiten durch Elektron-Phonon-Wechselwirkungsprozesse bestimmt werden. Ein sehr viel versprechendes Material ist hierbei der Hochtemperatursupraleiter YBCO, da dessen intrinsische Elektron-Phonon-Wechselwirkungszeit im einstelligen Pikosekundenbereich liegt. Bereits in den 90er Jahren wurde mittels elektro-optischem Sampling oder Pump-Probe-Experimenten die Elektronenrelaxation bei optischer Anregung mittels Femtosekundenlasern zu unter 3 ps bestimmt. Diese Studien konnten jedoch nur bis zu Wellenlängen von ca. 10 μm durchgeführt werden, da oberhalb dieser Wellenlängen, im Besonderen im THz-Frequenzbereich, keine brillanten Pulsquellen mit Pulsdauern im Pikosekundenbereich zur Verfügung standen.

Erst durch die Entwicklung spezieller Betriebsmodi an Teilchenbeschleunigern, wie dem sogenannten low-alpha Modus an Elektronenspeicherringen, konnte die langwährende Lücke im Bereich der hochbrillanten, gepulsten Terahertz-Quellen in einem Frequenzband von 0.1 - 2 THz geschlossen werden. Durch die Reduktion der Elektronenpaketlänge ist es möglich, kohärente THz-Strahlung mit sehr hoher Intensität und Pulsdauern im Bereich von wenigen Pikosekunden zu erzeugen. Um diese Strahlung zu analysieren und den low-alpha Betriebsmodus zu optimieren, sind ultra-schnelle Terahertz-Detektoren erforderlich, welche direkt im Zeitbereich dynamische Vorgänge im Pikosekundenbereich auflösen können. Aufgrund der Breitbandigkeit der emittierten Strahlung von 0.1 - 2 THz ist die Einbettung des Detektorelements in eine breitbandige THz-Planarantenne erforderlich. Da auch der emittierte Leistungsbereich in Elektronenspeicherringen über einen großen Bereich variiert werden kann, ist ein breiter dynamischer Detektionsbereich eine weitere Anforderung an die zu entwickelnde Detektortechnologie.

Diese Anforderungen werden von YBCO THz-Direktdetektoren erfüllt, wie in der vorliegenden Arbeit erläutert wird.

Das Ziel der vorliegenden Arbeit bestand darin, extrem schnelle Detektoren basierend auf dem Hochtemperatursupraleiter YBCO zu entwerfen, zu charakterisieren und optimieren und zuletzt in ein Detektionssystem mit Pikosekunden-Zeitauflösung zu integrieren, welches zur Charakterisierung gepulster THz-Quellen eingesetzt werden kann.

Dazu wurde zunächst die YBCO-Dünnschichtabscheidung mittels Laserablation optimiert. Der Fokus hierbei lag auf der Reduktion der Schichtdicke, bei gleichzeitig hoher supraleitender Qualität der Dünnschichten, für die Entwicklung schneller und sensitiver Detektoren. Durch die Einführung zweier Pufferschichten sowie einer Schutzschicht, konnten langzeitstabile YBCO Dünnschichten von nur 10 nm mit einer Sprungtemperatur von 79 K erfolgreich abgeschieden werden, was aufgrund der Höhe der YBCO Einheitszelle nur 8 Einheitszellenlagen entspricht. Damit wurde die Kühlung der YBCO Detektoren auch bei dünnsten Schichten von nur 10 nm mit flüssigem Stickstoff oder kompakten Kleinkühlern sichergestellt, wodurch die Verwendung von kostspieligem Helium, welches für viele andere THz-Direktdetektoren benötigt wird, obsolet wird.

Zur Entwicklung von schnellen und sensitiven YBCO-Detektoren war die Entwicklung eines Strukturierungsprozesses mit lateralen Abmessungen im Mikrometer- und Submikrometerbereich erforderlich. Dies erfolgte durch den Einsatz von Elektronenstrahllithographie und deren detaillierte Optimierung. Die Herausforderung bestand hierbei in der Entwicklung eines Strukturierungs- und Ätzprozesses für Submikrometer-Detektorelemente aus YBCO mit hoher supraleitender Qualität. Da die supraleitenden Eigenschaften von YBCO im Wesentlichen durch den Sauerstoffgehalt im Material bestimmt werden, ist es notwendig einen Ätzprozess zu entwickeln, welcher die Stöchiometrie des Materials nicht angreift. Dies wurde durch einen mehrstufigen Ätzprozess mittels physikalischem Ionenstrahlätzen und chemischem Nassätzen erreicht und Detektorelemente mit Längen von nur 300 nm wurden erfolgreich hergestellt. Alle Detektoren in dieser Arbeit weisen mit einer kritischen Temperatur oberhalb von 83 K eine sehr gute supraleitende Qualität auf. Ebenso wurde eine Langzeitstabilitätsstudie der supraleitenden Eigenschaften der Detektoren durchgeführt und gezeigt, dass über einen Zeitraum von über einem Jahr die kritische Temperatur und kritische Stromdichte unverändert gut bleiben.

Die entwickelten Detektoren wurden mittels optischer und THz-Strahlung charakterisiert. Zunächst wurde mittels optischer Strahlung die bereits früher erforschten Zeitkonstanten für die Energierelaxation in YBCO Dünnschichten für Photonenanregungen oberhalb der supraleitenden Energielücke überprüft. Es wurde eine Elektron-Phonon-Wechselwirkungszeit von 3.9 ps erzielt, was gut mit früheren Studien übereinstimmt. Zudem wurde eine schichtdickenabhängige Phononrelaxation im Nanosekundenbereich ins Substrat bestimmt, was ebenfalls in Übereinstimmung mit früheren Werten ist. Dieses typische Verhalten von YBCO-Dünnschichten bei Anregung von kurzen Pulsen mittels Strahlungsenergien oberhalb der supraleitenden Energielücke wird durch das Zwei-Temperaturen-Modell beschrieben. Der grundlegende Gedanke hierbei ist die Absorption der Strahlung im Elektronensystem des Supraleiters, was aufgrund der hohen Energie der absorbierten Strahlung zum Aufbrechen von Cooper-Paaren und somit "heißen" Elektronen führt (daher Hot-Electron-Bolometer). Das Elektronensystem relaxiert durch Abgabe der Energie an das Phononensystem im Supraleiter, wobei im letzten Schritt die Energie in das Substrat abgeführt wird.

Erst durch die Entwicklung des low-alpha Betriebsmodus an Elektronenspeicherringen, konnten die Studien der YBCO Strahlungsabsorption und anschließender Energierelaxation in dieser Arbeit auf Photonenenergien unterhalb der supraleitenden Energielücke ausgeweitet werden. Die entwickelten Detektoren wurden nach der Analyse mittels optischer Strahlung oberhalb der Energielücke, mit gepulster kohärenter THz-Strahlung unterhalb der YBCO Energielücke bestrahlt. Hierbei konnte erstmals gezeigt werden, dass der Detektionsmechanismus im THz-Frequenzbereich für gepulste Anregungen grundlegend verschieden zum optischen Wellenlängenbereich ist. Die Phononrelaxation im Nanosekundenbereich, welche kennzeichnend für das Zwei-Temperaturen-Modell ist, konnte unter keinen Umständen im THz-Frequenzbereich erzeugt werden. Noch eindeutiger war der Unterschied in der Detektorantwort ohne angelegten Biasstrom. Da der Detektionsmechanismus im optischen Wellenlängenbereich auf einer elektronischen Erwärmung und somit einer Widerstandsänderung basiert, ist ohne angelegten Strom keine Detektorspannung messbar. Im THz-Frequenzbereich konnte jedoch ohne jegliche Biasversorgung eine klare Detektorantwort gemessen werden. Diese sowie weitere in dieser Arbeit diskutierten Indizien legten nahe, dass im THz-Frequenzbereich bei Anregungen unterhalb der Energielücke ein anderer Detektionsmechanismus zugrunde liegt.

Ein mögliches Modell zur Erklärung der gefundenen experimentellen Ergebnisse basiert auf einem Detektionsmechanismus, der die Anregung magnetischer Flussschläuche in Betracht zieht. Die grundlegende Idee hierbei ist, dass der durch die einfallende Welle in der Antenne erzeugte HF-Strom eine Lorentzkraft im Detektorelement erzeugt, welche zur dissipativen Bewegung der Flussschläuche führt und somit die Detektorantwort erzeugt. Dies bedeutet, dass die Detektorantwort nicht durch die Intensität der einfallenden Strahlung bestimmt wird, wie das bei optischen Anregungen der Fall ist, sondern vielmehr durch das elektrische Feld der einfallenden Welle. Diese Hypothese konnte im Rahmen der Arbeit bewiesen werden, indem der verwendete Spiegel im Strahlengang variiert wurde. Durch das Austauschen des metallischen Spiegels durch einen dielektrischen Spiegel wurde die Phase des elektrischen Feldes der einfallende Welle um 180 Grad gedreht. Dies führte direkt zu einer Umkehrung der Detektorantwort, was eindeutig zeigt, dass der Detektor nicht auf die Intensität sondern auf das elektrische Feld der einfallenden Welle reagiert.

Durch das Fehlen der Nanosekundenkomponente in der THz-Detektorantwort erzielten die entwickelten Detektoren ultra-schnelle Antwortzeiten im Pikosekundenbereich. Um diese zu bestimmen und Pikosekundenprozesse im THz-Frequenzbereich analysieren zu können, wurden die Detektoren in ein extrem breitbandiges und somit schnelles Auslesesystem integriert. Hierfür wurde ein neues Detektionssystem am IMS aufgebaut, welches von der quasioptischen Einkopplung der THz-Strahlung über die Ausleseelektronik bis hin zu den kryogenen Komponenten optimiert wurde. Die effektive Auslesebandbreite des entwickelten Systems erlaubt zeitliche Prozesse von 15 ps (Halbwertsbreite) aufzulösen, wodurch es erstmals in dieser Arbeit möglich wurde, die Zeitstruktur kohärenter THz-Pulse erzeugt am Elektronenspeicherring ANKA des KIT in Echtzeit aufzulösen. Es wurde eine Halbwertsbreite von 17 ps bestimmt, was sehr gut mit theoretischen Berechnungen, unter Berücksichtigung der verwendeten Beschleunigereinstellungen, übereinstimmt. Mit dem entwickelten YBCO Detektionssystem wurde zudem das Burstingverhalten der emittierten THz-Strahlung an der Ultraviolet Synchrotron Radiation Facility (UVSOR) in Japan studiert. Weitere Experimente wurden zur Analyse des Füllmusters der gespeicherten Elektronenpakete am Freien-Elektronen-Laser der Universität Osaka in Japan durchgeführt sowie das dynamische Verhalten bei der Emission von Quantenkaskadenlaser in Leeds analysiert.

Zusammenfassend kann gesagt werden, dass die Experimente mit den in dieser Arbeit entwickelten THz-Detektoren basierend auf dem Hochtemperatursupraleiter YBCO wichtige theoretische Fragestellungen im Bereich des Detektionsmechanismus in YBCO Dünnschichten für Anregungen unterhalb der Energielücke beantworten konnten. Zudem wurde ein Direktdetektionssystem für den Terahertzfrequenzbereich entwickelt und aufgebaut, welches mit einer zeitlichen Auflösung von 15 ps neue Möglichkeiten bei der Analyse von dynamischen Prozessen im Zeitbereich eröffnet, wie zum Beispiel im Bereich der Beschleunigerphysik.

Contents

1. Introduction

Infrared radiation up to terahertz (THz) wavelengths is extensively used in many different research areas such as spectroscopy, wireless communication and cosmology [1]. To date, many of these applications have been focused on high-resolution spectroscopy leading to the employment of heterodyne detectors which allow for ultra-sensitive detection near the quantum limit [2]. However, with the recent emergence of new THz sources which provide radiation with high brightness and power on ultra-short time scales down to single picoseconds, more emphasis is now being focused toward direct detection techniques and components [2]. Accelerator-based sources of radiation provide these desired properties over a very broad frequency range from X-ray up to THz wavelengths. Already in the 1980s, potential applications of pulsed-infrared synchrotron radiation were suggested. They are as diverse as the study of the spectra of liquid crystals and the use of far-infrared synchrotron radiation as transfer standard in calibrating radiometers [3].

In recent years, the development of electron storage rings emitting ultra-short, brilliant pulses in the THz frequency range made significant progress. An intrinsic property of an electron storage ring is the distribution of the electron beam current into bunches with a minimum distance corresponding to the accelerating radio frequency system. If the bunch length is much longer than the emitted radiation wavelengths, non-coherent broadband radiation from low THz frequencies up to the X-ray frequency range is emitted. For electron bunches which are in the order of and shorter than the radiation wavelength coherent synchrotron radiation (CSR) is emitted leading to a strong enhancement of the emitted power in the THz frequency range [1]. Using dedicated magnetic lattices, so-called low-alpha optics [4–6], is one possibility to produce broadband THz CSR (0.1 - 2 THz) pulses with pulse lengths in the single picosecond range [4, 7, 8]. The emitted CSR power of an electron storage ring scales with the stored electron beam current. The possibility to store any current from 1 pA (single electron) to over 100 mA in the ring [7] leads to the fact that not only the emitted radiation power can be tuned over a very wide range but also the absolute emitted THz power level is very high. This makes electron storage rings the most powerful sources for pulsed THz radiation nowadays [7].

For the analysis and optimization of the pulsed THz radiation generated by electron storage rings or other pulsed sources, ultra-fast detectors are required which are able to resolve picosecond dynamic processes directly in the time domain. A promising candidate for the analysis of THz picosecond pulses are detectors made from the high-temperature superconductor $YBa_2Cu_3O_{7-x}$ (YBCO). Detailed studies on the energy relaxation processes in YBCO in the optical frequency range were already carried out [9–12]. By means of electro-optical sampling technique an electron energy relaxation time of only a few picoseconds was determined [11], which allows for ultra-fast time-domain measurements.

However, in all these experiments the photon energy of the incoming radiation was above the superconducting energy gap in YBCO, and hence the absorbed radiation resulted in a Cooper pair breaking with subsequent electron and phonon cooling processes.

The goal of this work is to transfer the experience from the optical wavelengths to the THz frequency range, to analyze the influence of sub-energy gap excitations on the detection mechanism and to develop an ultra-fast YBCO direct detector for the THz frequency range. With respect to one major application, the analysis of coherent synchrotron radiation, the requirements for the new direct detector technology are: ultra-fast temporal resolution on a picosecond time scale, a broad THz frequency detection range and last but not least high responsivity and sensitivity values together with a broad detection dynamic range.

To motivate the use of the high-temperature superconductor YBCO as direct detector material, a review of the state of the art of direct THz detector technologies is given in chapter 2. Whereas in section 2.1 the most famous representative of direct THz detectors, the resistive bolometer, is discussed, other well-established direct THz detectors are summarized in section 2.2. In section 2.3 the characteristics of the state of the art direct detectors are finally summarized together with the requirements of picosecond THz pulse detection over a broad frequency and power range demonstrating the need to develop a new direct THz detector technology based on YBCO.

For the optimization of a detector technology a clear understanding of the mechanism responsible for radiation detection is required. One major influence on the responsible detection mechanism in superconducting materials is the incident photon energy which may be above or below the energy gap of the superconductor. Therefore, in chapter 3 the detection mechanism for over-gap and sub-gap excitations are discussed. Before the discussion of the existing models, the high-temperature superconductor YBCO with its properties influencing the radiation detection is reviewed in section 3.1. The well-studied models in the optical frequency range, i.e. for over-gap excitations, are summarized in section 3.2. For the THz frequency range, i.e. for sub-gap excitations, only very few reports exist up to now which are discussed in section 3.3.

For the development of the YBCO THz detectors the pulsed-laser deposition process for thin YBCO films of 10 to 100 nm thickness has been optimized at the IMS (chapter 4). The YBCO thin films were deposited on sapphire substrate and embedded in a multi-layer structure required for the detector development.

After the optimization of the as-deposited films, a patterning technology for micrometer- and sub-micrometer-sized detecting elements embedded in a planar THz antenna has been developed as described in section 5.1. The DC characterization of the patterned YBCO detectors as well as a long-term study of the superconducting characteristics of the devices is discussed in section 5.2. In chapter 6 detailed experiments concerning the photoresponse of the developed detectors in the optical and THz frequency range have been carried out and analyzed to get insight in the detection mechanism in the respective frequency ranges. Significant differences between over-gap (section 6.1) and sub-gap (section 6.2) excitations in YBCO thin-film detectors were found (section 6.3) leading to a new understanding of the detection mechanism in the THz frequency range for pulsed sub-gap excitations discussed in section 6.4.

In chapter 7 the application of the developed YBCO thin-film detectors at several pulsed THz sources is demonstrated. For this, the YBCO thin-film detectors were embedded in an ultra-fast detecting system which was developed during this work with readout electronics up to 65 GHz (section 7.1). The successful measurements at several electron storage rings are discussed in section 7.2.1. The first measurement of the real-time evolution of a THz CSR pulse at ANKA is shown and the successful measurements of bursting CSR at the electron storage ring UVSOR-II in Japan are discussed. Further successful measurements at a free-electron laser (section 7.2.2) and a quantum-cascade laser (section 7.2.3) demonstrate the suitability of our YBCO detector system for the analysis of ultra-fast dynamic processes in the THz frequency range on a picosecond time scale.

2. Direct THz detectors - State of the art

Terahertz (THz) radiation comprises electromagnetic waves propagating at frequencies from 0.3 to 3 THz (wavelength range of 1000 to 100 μm) [2]. Detection at THz wavelengths differs from detection at shorter optical wavelengths and longer radio wavelengths.

In comparison to shorter wavelengths, the photon energies levels at THz wavelengths are low ranging from 1.2 to 12.4 meV or to an equivalent black body temperature of 14 to 140 K, well below the ambient background on earth (thermal energy of 26 meV at room temperature). Thus, ambient background thermal noise almost always dominates naturally emitted narrow-band signals requiring either cryogenic cooling or long-integration-time techniques or both. Besides, the Airy disk diameter (diffraction limit) is rather large (hundreds of micrometers) that makes a matched director, i.e. an antenna between the signal and the detector element necessary [2].

In comparison to longer wavelengths, THz technology suffers from a lack of electronic components: lumped resistors, capacitors and inductors, as well as amplifiers and low-loss transmission lines are not available to THz detectors [2].

Direct THz detectors are the medium of choice for applications that do not require ultra-high spectral resolution. Typical direct detector technologies in use are the following [13]:

- composite bolometers: the thermometer is attached to a separate absorber [14],

- monolithic bolometers: the thermometer and absorber are identical [14] with the particular case of microbolometers for longer i. e. THz wavelengths, where an antenna is used to couple power to a small thermally absorbing region,

- Schottky-barrier diodes: used as square-law detectors, and

- Golay cells: based on thermal absorption in a gas-filled chamber and a detected change in volume via a displaced mirror in an optical amplifier.

Some traditional infrared (IR) detectors respond also in the THz frequency range [13]. These detectors are

- pyroelectric detectors: its dielectric constant changes as a function of temperature, and

- photoconductors: based on mechanically stressed gallium-doped germanium or HgCdTe.

In section 2.1, the widely used bolometer concept will be discussed in detail. The state of the art of this detector technology will be compared to the other prominent direct THz detector technologies listed above (section 2.2).

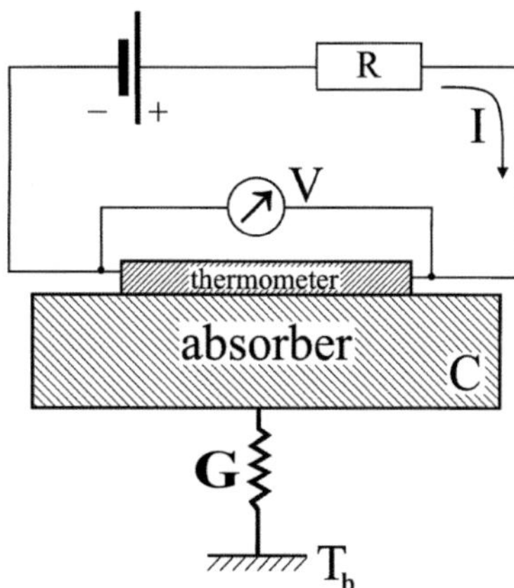

Fig. 2.1.: A composite bolometer consists in general of an absorber characterized by its specific heat capacity C, a resistive thermometer and a thermal link with the thermal conductance G to a constant-temperature reservoir T_b. Radiation is absorbed by the absorber which changes its temperature. The thermometer responds to changes of temperature by a change of resistance. It is measured e.g. by passing a constant current I through the thermometer and by observing the voltage drop across it [14].

The different detector technologies will be evaluated in terms of their suitability for

- fast time-domain measurements on a picosecond time scale

- over a wide THz frequency range (0.1 - 2 THz) and

- a wide power range.

In section 2.3 this review of direct THz detectors is summarized and evaluated demonstrating the need for the development of a new direct THz detector technology for ultra-fast temporal processes with a wide dynamic range over a broad frequency range.

2.1. Resistive bolometer detectors

The noun 'bolometer' is a composite word of Greek origin, namely of bole (beam, ray) and metron (meter, measure). A resistive bolometer is a type of thermal detector where the electrical resistance of the material is the property that is measured in response to incident electromagnetic radiation [13]. The active element may be a metal, a semiconductor or a superconductor which converts a temperature rise into a change of electrical resistance.

A bolometer consists of an absorber for radiation and a temperature transducer (thermometer) which are attached to a volume having a heat capacity $C(T)$, which is linked by a thermal conductance $G(T)$ to a heat sink or stable temperature bath at a temperature T_b (see Fig. 2.1) [14]. Bolometers where a thermometer is attached to a separate absorber are referred to as composite bolometers; those where the thermometer and the absorber are identical are called monolithic detectors. For fast and sensitive detectors the volume of the detecting element has to be reduced to micrometer or even sub-micrometer dimensions which is below THz wavelengths. To couple THz radiation to these sub-wavelengths devices antennas are used. These kind of detectors are called microbolometers.

2.1.1. Figures of merit

In practice, a bolometer measures the amount of radiation power incident on an active area by producing a corresponding electrical signal. Initially the entire apparatus is at thermal equilibrium at bath temperature T_b. Radiant power $P(t)$ is incident upon the sensitive element, whose absorptivity is η and which is biased with a constant current I. If radiation is absorbed, the temperature increases to T, giving rise to a detector signal. The temperature response of the bolometer according to Fig. 2.1 with the heat capacity C and the thermal conductance G can be calculated according to the differential equation for the heat flow in the detector [14]:

$$C\frac{dT}{dt} + G(T - T_b) = \eta P(t) + RI^2. \tag{2.1}$$

For a first-order expansion of the thermometer resistance R of $R = R_0 + (dR/dT)\Delta T$, the equation for the time dependent part of the temperature change $\Delta T = T - T_b$ becomes according to [14]

$$C\frac{d(\Delta T)}{dt} + G_{eff}\Delta T = \eta P(t) \tag{2.2}$$

with the effective thermal conductance G_{eff}

$$G_{eff} = G - I^2\frac{dR}{dT} = G - R_0 I^2 \alpha \tag{2.3}$$

where α, the relative temperature coefficient of resistance, is given by

$$\alpha = \frac{1}{R_0}\frac{dR}{dT}. \tag{2.4}$$

The effective thermal conductance, G_{eff}, is a modification of the thermal conductance, G, due to the electrothermal effect. This effect comes about because the bias power dissipated in the bolometer changes when the received radiation causes a change in the bolometer resistance. If the bolometer temperature coefficient of resistance, α, is positive as e.g. for superconductors and the bias is supplied by a constant current source, the received radiation will cause the bolometer resistance to increase and more bias power will be dissipated in the bolometer. The net effect of this will be to compensate somewhat for the heat loss due to the thermal conductance, G. This positive electrothermal feedback results in an effective thermal conductance, G_{eff}, that is less than G. If the bias current is not limited, the radiation may cause a continuous increase in the bolometer bias power until G_{eff} becomes less than zero and the bolometer is destroyed by "thermal runaway" [15]. If the bias current is supplied by a constant voltage source and α is positive, the radiation will cause the bias current to decrease and less bias power will be dissipated in the bolometer. The net effect of this negative electrothermal feedback will be that G_{eff} will be greater than G [15]. If α is negative, the electrothermal feedback will be positive if the bias is from a constant voltage source, and negative if it is from a constant current source.

The solution of the bolometer temperature response (equation 2.2) in the frequency domain is according to [14]

$$\Delta T = \frac{\eta P(\omega)}{(G_{eff}^2 + \omega^2 C^2)^{1/2}} \tag{2.5}$$

with the absorbed power $P(\omega)$ in the bolometer, modulated at an angular frequency ω.

The responsivity of a bolometer is defined as $S = \Delta V/P$ [14], where ΔV is the voltage drop across the thermometer as a consequence of a change in power input P:

$$S = \frac{\Delta V}{P} = \frac{\eta \alpha R_0 I}{(G_{eff}^2 + \omega^2 C^2)^{1/2}}. \tag{2.6}$$

With the thermal time constant τ of a bolometer given by

$$\tau = \frac{C}{G_{eff}} \tag{2.7}$$

the responsivity results in

$$S = \frac{\eta \alpha R_0 I}{G_{eff}(1 + \omega^2 \tau^2)^{1/2}} \tag{2.8}$$

Thus the responsivity is frequency-independent at low modulation frequencies and rolls off with frequency at higher frequencies. The -3 dB value ($0.707 S_0$, where S_0 is the zero-frequency responsivity) is determined by the condition $\omega \tau = 1$. Equation 2.8 shows that the key to developing highly sensitive bolometers is having a high temperature coefficient of resistance α, a very low thermal heat capacity C and excellent thermal isolation (low thermal conductance G). This directly leads to the consequence that highly sensitive detectors with small thermal conductances always show large response times (see equation 2.7).

The noise equivalent power (*NEP*) of a detector is defined as the input power required such that the output signal-to-noise ratio equals unity. To say it different: The *NEP* of a bolometer is defined as the power in a 1 Hz bandwidth one has to present to the detector in order to receive a response of the same signal height as the noise [14]. If the bolometer bias is supplied from a constant current source, the *NEP* of the bolometer detector can be expressed as

$$NEP = \frac{V_n}{S} \tag{2.9}$$

where

V_n = rms noise voltage, and

S = responsivity of the bolometer.

V_n is often normalized with respect to the square root of the detector video (or post-detection) bandwidth, and the *NEP* is then expressed in units of $W \cdot Hz^{-0.5}$.

2.1.2. Composite bolometers

Composite bolometers employ separate absorber and thermometer. This allows for the development and optimization of each part separately which makes it possible to keep the bolometer design flexible. The absorbing material is chosen to effectively capture the incoming radiation (and so must have an area of about λ^2 and a sheet resistance equal to the free-space impedance $Z_0 = 120\pi\ \Omega$), but can be relatively thin to give a low thermal heat capacity [16]. The temperature sensor material is chosen to produce very large changes in resistance with changing temperature, and must be in intimate thermal contact with the absorber, but may otherwise be very small and highly resistive. This composite structure can have an overall thermal heat capacity which is much less than for monolithic bolometers.

Semiconducting thermometers

A typical composite bolometer consists of bismuth (Bi) as the radiation absorber and gallium-doped germanium as the temperature sensor [17]. The review article by Richards [18] discusses in detail composite semiconductor detectors operating at or below the temperature of liquid helium. For applications with very low background radiation (space platform) at 300 mK electrical sensitivity values of $8.7 \cdot 10^7$ V/W and *NEP* values of $1 \cdot 10^{-16}$ $WHz^{-0.5}$ at a time constant of 11 ms were achieved [19]. Micromesh neutron transmutation doped Ge bolometers developed for the HFI Planck/Herschel mission operated at 100 mK achieved *NEP* values of $1 - 10 \cdot 10^{-17}$ $WHz^{-0.5}$ with time constants between 1 and 10 ms [20].

Commercially available is the semiconducting Ge bolometer from QMC Instruments Ltd. This composite bolometer operates at cryogenic temperatures near 4.2 K. It incorporates a Germanium thermistor attached to a metallic absorber which is in turn deposited onto a thin silicon nitride support substrate [21]. The detector's electrical responsivity at 4.2 K amounts to $2 \cdot 10^4$ V/W with a *NEP* of $1 \cdot 10^{-12}$ $WHz^{-0.5}$ at 80 Hz. These detectors can be employed over a very wide frequency range from 0.09 up to 30 THz. However, their response times are very slow on the millisecond time scale [21]. Very recently, QMC Instruments Ltd. replaced the Germanium bolometer by a low-T_c Nb superconducting composite bolometer [21].

Composite bolometers with semiconducting thermometers encompass a broad frequency detection range and are ultra-sensitive devices with large responsivity values. However, their response times are rather large with time constants on the millisecond scale making these devices not suitable for picosecond time-domain measurements.

Low-T_c superconducting thermometers

Research on thermal detectors for astronomical and spectroscopic investigations showed that sensitive detectors could be developed using materials that became superconducting when cooled to liquid helium temperatures. These materials have a high temperature derivative of resistance on the transition from normal to superconducting state. This results in very sensitive bolometers when operated around their superconducting transition temperature [14, 18]. Furthermore, since the specific heat of most materials decreases significantly with temperature, the time constant shortens and thus faster detectors as compared to room-temperature detectors can be built. Research on superconducting bolometers began in 1938 at the Johns Hopkins University [22]. This work was motivated by the idea that exploitation of superconductivity could lead to faster and more sensitive bolometers. The results of this work dealing with a composite tantalum bolometer were published in 1942 [23]. Independently, Goetz suggested superconducting bolometers [24]. The development of low-T_c superconducting composite bolometers was reviewed by Clarke *et al.* [25]. They fabricated a composite structure of a Bi absorbing film on the reverse side of the substrate and an aluminum transition-edge bolometer. A *NEP* of $1.7 \cdot 10^{-15}$ WHz$^{-0.5}$ and a responsivity of $7.4 \cdot 10^4$ V/W at 2 Hz at an operating temperature of 1.27 K with a time constant of 80 ms was achieved. They also discussed the possibility to use a superconductor-normal metal-superconductor Josephson junction or a superconductor-insulator-normal metal quasiparticle tunneling junction as thermometer [25]. Nowadays, low-T_c transition edge sensor bolometers developed for radioastronomy applications operated below 1 K achieve ultra-low *NEP* values. Membrane-isolated Mo/AuPd bolometers operated at 300 mK achieved *NEP* values of $8 \cdot 10^{-16}$ WHz$^{-0.5}$ [26]. *NEP* values as small as $5 \cdot 10^{-20}$ WHz$^{-0.5}$ are expected for Mo/Au transition-edge sensor bolometers with Si$_3$N$_4$ isolation beams operated at 60 mK developed for the SPICA/BLISS mission [27].

Composite bolometers made from low-T_c superconducting thermometers are comparable in their performance to semiconducting composite bolometers. They are very sensitive and show large responsivity values. However, with response times on the millisecond time scale these devices can not be employed for direct time-domain measurements on a picosecond time scale.

High-T_c superconducting thermometers

Detectors which operate at about liquid nitrogen temperatures are much more acceptable for many applications than those that require being cooled to liquid helium temperatures, especially since relatively low-cost closed-cycle refrigerators (cryocoolers) are available. Furthermore, the belief that high-T_c superconductors had a energy gap which was much larger than in metallic superconductors yet smaller than for most semiconductors raised the possibility that very-long-wavelength photon detection at operating temperatures near that of liquid nitrogen was possible [28]. This raised the hope that their exploitation as infrared detectors might provide performance advantages over conventional infrared detectors based upon semiconductors.

Verghese *et al.* [29] fabricated a composite high-T_c $YBa_2Cu_3O_{7-x}$ (YBCO) bolometer with a *NEP* of $2.4 \cdot 10^{-11}$ $WHz^{-0.5}$ at 10 Hz, a responsivity of 17 V/W and a time constant of 55 ms. Gold black smoke was deposited on the backside of the assembled bolometer as an absorber. Spectral measurements revealed that the bolometer had a reasonable sensitivity from visible wavelengths to beyond 100 μm [29]. Brasunas *et al.* [30] demonstrated a *NEP* of $1.6 \cdot 10^{-11}$ $WHz^{-0.5}$ for a composite YBCO transition-edge bolometer on a thinned sapphire substrate which was a record value at this time. However, the time constant was as large as 300 ms. A high-T_c GdBaCuO bolometer on a silicon nitride membrane was developed by De Nivelle *et al.* [31]. A gold black absorption layer was used to couple radiation from 70 to 200 μm. A responsivity of 6900 V/W with a *NEP* of $3.8 \cdot 10^{-12}$ $WHz^{-0.5}$ at 2 Hz and a time constant of 115 ms were achieved.

Composite bolometers with high-T_c superconducting thermometers allowed to extend the photon detection to longer wavelength compared to semiconducting composite bolometers and to operate the detector at liquid nitrogen temperatures at the cost of low *NEP* values. However, also for these devices millisecond response times were observed which do not allow to resolve picosecond dynamic processes.

Summary

Composite bolometers, in particular made from semiconducting and low-T_c thermometers, show high responsivity values of 10^7 V/W and ultra-high sensitivities well below 10^{-16} $WHz^{-0.5}$. However, with regard to a picosecond temporal analysis in the THz frequency range this bolometer-type is unsuitable due to very large response times on the millisecond time scale.

2.1.3. Monolithic bolometers - Direct absorbers for infrared wavelengths

For a monolithic bolometer the absorber and thermometer are identical. In order to directly absorb incident radiation the detector must be comparable in size to the wavelength of the radiation and be impedance-matched to free space [16]. Monolithic bolometers as direct absorbers for infrared wavelengths are typically several wavelengths square, which may be up to several square millimeters for THz radiation. Thus, a drawback of these monolithic-type bolometers is their large thermal heat capacity, which reduces the speed at which they can respond to temperature changes [16]. Also, the *NEP*, which should be as small as possible for sensitive operation, increases approximately as the square root of the bolometer area [16].

Normal metal monolithic bolometers

The original bolometer was constructed by Langley, who used platinum foils and ribbons for solar observations [32]. Block *et al.* [33] transferred the bolometer concept to infrared wavelengths by studying thin-film metal room-temperature detectors at 10.6 μm. Due to the low thermal conductivity of Bi, these films achieved higher responsivities (up to 22 mV/W with α = -0.35 %/K) compared to nickel, antimony and chromium which is in agreement with equation 2.8. Although Bi has therefore longer relaxation time constants, response times as short as 2 ns were measured. Rebeiz *et al.* [34] developed large-area Bi monolithic detectors for absolute power calibration applications. They demonstrated a 1x1 cm^2 room-temperature Bi bolometer with a responsivity of 1 mV/W at 100 Hz. The *NEP* of this device was determined to $3 \cdot 10^{-6}$ WHz$^{-0.5}$ at 1 kHz. A monolithic bolometer made from normal metal was the first bolometer. However, nowadays, particular in the THz frequency range, these detectors are replaced by semiconducting monolithic bolometers which are discussed in the following.

Semiconductor monolithic bolometers

Considerable effort has gone into the development of room-temperature bolometers that have sufficient sensitivity for many infrared imaging applications. This effort has led to the development of materials that have a relatively high temperature coefficient of resistance. Prominent amongst these materials are vanadium oxide (VO$_x$), amorphous silicon (α-Si) and yttrium barium copper oxide (YBCO). These materials are semiconductors at room temperature with temperature coefficients of resistance of -2%/K for VO$_x$, -2.5%/K for α-Si and higher than -3%/K for semiconducting YBCO [35]. Room-temperature bolometer focal plane arrays (FPA) at infrared wavelengths (λ = 8 - 14 μm) were successfully developed for all three materials [36–38].

Very sensitive bolometers can be made with cooled semiconductor thermometers. These bolometers have a long history in infrared detection. F. J. Low used a gallium doped germanium single crystal and operated it at a temperature of T = 2 K [39]. He found a *NEP* of $5 \cdot 10^{-13}$ WHz$^{-0.5}$ and a time constant of 400 μs. A temperature coefficient α of -200 %/K was achieved.

Since the days of these early semiconductor thermometers, improvements have been made and values of α between -400 %/K and -1000 %/K can be produced at low temperatures [14].

However, not only very sensitive but also fast devices (microsecond time scale) can be fabricated from semiconducting bolometers using the hot-electron effect. One example is the commercially available Indium antimonide (InSb) bolometer from QMC Instruments Ltd [21] operated at 4.2 K. According to [21] at liquid helium temperatures intra-band free-electron absorption of rather long wavelengths results in changes of the electron mobility. The electron-electron interaction time is orders of magnitude shorter than that of the electron-phonon interaction. The electrons therefore come into thermal equilibrium at a temperature above the lattice. This temperature change and the resulting change in electron mobility, termed the hot-electron response, allows response times in the microsecond range. This is much faster than other semiconducting counterparts. The InSb hot-electron bolometers show responsivity values of 5000 V/W with a *NEP* of $5 \cdot 10^{-13}$ WHz$^{-0.5}$ at kHz modulation frequencies. However, the useful frequency range is quite narrow between 0.06 and 0.5 THz [21].

Semiconducting monolithic bolometers are sensitive and highly responsive devices. In general, the temperature dependence of the resistance of a doped semiconductor, usually Si or Ge, is the most widely used material for bolometers operated at or below liquid helium temperatures [18]. However, the bulk absorption coefficient of Ge and Si thermometer material with a useful resistivity decreases at low frequencies, since the photon energy $h\nu$ becomes small [18]. Consequently, bolometers for millimeter wavelengths must not only have large area, but must typically be one or more millimeters thick. The resulting heat capacity is a significant limitation [18] leading to large time constants. The fastest semiconducting monolithic bolometers are based on the hot-electron effect as discussed above and reach time constants on the microsecond time scale. This is still far too slow for the analysis of picosecond time-domain processes.

High-T_c YBCO superconducting monolithic bolometers

Also high-T_c superconductors were used for the development of monolithic bolometers as direct absorber. Oppenheim *et al.* [40] fabricated 120×200 μm^2 bolometers from epitaxial 1 μm thick high-T_c YBCO films on $SrTiO_3$ substrate. The responsivity of the bolometer at $\lambda = 10.6$ μm was determined to be 0.5 V/W ($f_m = 130$ Hz) which was close to the calculated value of 0.78 V/W. The response time was 11.3 μs. This result lies between a very fast high-T_c YBCO monolithic bolometer with low responsivity ($\tau = 4$ ns, $S = 0.018$ V/W) [41] and a slow YBCO composite bolometer with high responsivity ($\tau = 32$ s, $S = 5.2$ V/W) [42]. In [41] 200×600 μm^2 bridges with a thickness of 48 nm deposited on MgO and $LaAlO_3$ substrates were used and characterized with infrared pulsed synchrotron radiation. In [42] a 20 μm wide and 76000 μm long meander was used with a Bi film for absorption. The meander was characterized with a 500 K blackbody source. Reports on high-T_c YBCO superconducting monolithic bolometers are restricted to infrared wavelengths. Due to the low responsivity values this bolometer-configuration is not interesting for sensitive radiation detector applications. However, these first reports concerning nanosecond response times show the potential for fast time-domain analysis with this superconducting material.

Summary

Monolithic bolometers used as direct absorbers offer sensitive detection for infrared wavelengths at moderate response time (micro- and nanosecond time scale). However, for THz wavelengths these devices based on direct absorption are not suitable due to the reduced photon energies at these long wavelengths. Thus, antenna-coupled microbolometers need to be employed which are discussed in the following.

2.1.4. Monolithic bolometers - Antenna-coupled microbolometers for THz wavelengths

One major problem for the implementation of wide-band THz detectors is to couple the incident radiation to the active detector region efficiently [43]. For instance, YBCO exhibits very high reflectivity of more than 98% for wavelengths above 20 μm [44]. Since absorbing layers tend to hamper the device time constant, antenna coupling is preferred at sub-millimeter wavelengths [43].

Like the monolithic bolometer, an antenna is comparable in size to the wavelength, but unlike the monolithic bolometer, it dissipates no power. Instead, the antenna couples the power into the very small load resistor, which is then practically the only part of the system to change temperature. A small device like this also has a large thermal impedance which results in relatively large change in temperature for a given amount of dissipated power. If the load resistor is the bolometer itself, extremely fast and sensitive performance can result [16]. Such a detector is called a microbolometer [45, 46]. Antenna-coupled microbolometers have been shown to have higher responsivity, better sensitivity, and much faster response than other bolometer types [47]. Since these are thermal detectors, they work well throughout the THz spectral region without the capacitive roll-off associated with Schottky detectors (discussed in section 2.2.1).

Room-temperature antenna-coupled microbolometers

These devices exploit the electrical and thermal properties of metal microbridges, e.g. from bismuth or niobium [43]. The first realization was a 5×4 μm^2, 55 nm thick Bi microbridge deposited on a quartz substrate, with V-antenna coupling [45]. The *NEP* value was $1.6\cdot10^{-10}$ $WHz^{-0.5}$ in the far infrared. More than one order of magnitude improvement in the *NEP* value has been obtained with suspended microbridges or low thermal conductance buffer layers with silicon substrate. A detailed overview is given in Table 2.1. However, up to now none of these room-temperature technologies meet the requirement of a very low *NEP* for the use e.g. in radiometer applications [54]. Therefore, cooled detectors fabricated from superconducting materials are in focus of research.

Cooled antenna-coupled microbolometers

Conventional low-T_c superconductors can offer unsurpassed performances in terms of both voltage responsivity and *NEP* [43]. However, due to the effort of liquid helium cooling, only few realizations as direct detectors are reported up to date which are summarized in [43]. Low-T_c detectors are mainly used in mixing applications where high spectral resolution and ultimate sensitivity are required. A summary of low-T_c direct microbolometers for the THz frequency range is given in Table 2.2 showing that minimum response times of 165 ps were achieved for a NbN hot-electron bolometer limited by the employed readout electronics.

As high-T_c direct detectors, antenna-coupled YBCO bolometers have been proposed by Hu *et al.* and have been theoretically predicted to reach a phonon noise-limited *NEP* of $3\cdot10^{-12}$ $WHz^{-0.5}$ [61]. This is much lower than for other wide-band THz detectors like Golay cells or pyroelectric detectors. Nahum *et al.* implemented this concept by fabricating a $0.1\times6\times13$ μm^3 YBCO strip directly on a low thermal conductivity substrate of ZrO_2 stabilized Y_2O_3 [62]. A *NEP* of $4.5\cdot10^{-12}$ $WHz^{-0.5}$ at 10 kHz and a responsivity of 478 V/W at a bias current of 550 μA was found. The idea was further developed by Rice *et al.* [63] fabricating YBCO microbolometers on micromachined silicon. Above a thin Si window, a YSZ-CeO_2 buffer layer was grown on which subsequently a YBCO film was deposited. The section underneath the YSZ was etched away, thus

Table 2.1.: Characteristics of room-temperature millimeter- and sub-millimeter-wave microbolometer detectors.

Authors	Antenna/ configuration	Film/ substrate	S (V/W)	NEP ($pWHz^{-0.5}$)	f_m (Hz)	τ (s)
Hwang et al. [45]	V antenna	Bi/ Quartz	10	160	-	$2 \cdot 10^{-6}$
Neikirk et al. [46]	V antenna/ suspended bridge	Bi	100	30	-	$1 \cdot 10^{-6}$
Shimizu et al. [48]	Bow-tie/ suspended bridge	Bi/ Si_3N_4	71	-	-	-
Grossman et al. [49]	Log-spiral	Nb/ Si	28	1	-	-
MacDonald et al. [50]	Log-spiral	Nb/ SiO_2/ Si	21	110	1000	$1.8 \cdot 10^{-7}$
Rahman et al. [51]	Dipole	Nb/ Si_3N_4/ Si	-	83	1000	-
Luukanen et al. [52]	Log-spiral	Nb/ Polyimide/ Si	400	15	1000	$<1.5 \cdot 10^{-6}$
Cherednichenko et al. [53]	Log-spiral	YBCO/ CeO_2/ sapphire	15	450	5000	$2.5 \cdot 10^{-9}$

Table 2.2.: Overview of published data on sub-millimeter wave superconducting bolometers with planar antenna coupling based on conventional low-T_c superconductors. *These values were determined optically not accounting for all optical losses.

Authors	Antenna/ configuration	Film/ substrate	T (K)	S	NEP ($pWHz^{-0.5}$)	f_m (Hz)	τ (s)
Nahum et al. [55]	-	Mo:Ge	0.1	10^9 V/W	$1 \cdot 10^{-6}$	-	$1 \cdot 10^{-6}$
Wentworth et al. [56]	Bow-tie	Pb	7.7	274 V/W	-	-	-
Luukanen et al. [57]	Spiral	Nb (air-bridge)/ Si	4.2	-1430 A/W	0.014	1000	$9 \cdot 10^{-7}$
Semenov et al. [58]	Spiral	NbN/ Si	4.2	5 V/W*	6000*	30	$1.65 \cdot 10^{-10}$
Zhang et al. [59]	Spiral	NbN/ Si	8	7600 A/W	0.3	2000	-
Santavicca et al. [60]	Double-slot	Nb/ Si	5.2	44000 V/W	0.02	>100	$0.7 \cdot 10^{-9}$

Table 2.3.: Overview of published data on sub-millimeter wave superconducting bolometers with planar antenna coupling based on high-T_c YBCO.

Authors	Antenna/ configuration	Film/ substrate	T (K)	S (V/W)	NEP (pWHz$^{-0.5}$)	f_m (Hz)	τ (s)
Hu et al. [61]	-	YBCO/ MgO	90	-	60	-	$6 \cdot 10^{-10}$
Hu et al. [61]	-	YBCO/ YSZ	90	-	5	-	$2 \cdot 10^{-7}$
Barholm-Hansen et al. [64]	Log-periodic	YBCO/ MgO	81	25	400	500	-
Nahum et al. [62]	Log-periodic	YBCO/ YSZ	91	480	4.5	500	$2 \cdot 10^{-5}$
Rice et al. [63]	Log-periodic/ suspended bridge	YBCO/ Si$_3$N$_4$	88	2900	9	-	$1 \cdot 10^{-5}$
Khrebtov et al. [65]	Bow-tie	YBCO/ NdGaO$_3$	85	300	12	-	$3 \cdot 10^{-7}$
Hammar et al. [66]	Log-spiral	YBCO/ CeO$_2$/ sapphire	77	190	20	500 - 100000	$3 \cdot 10^{-10}$

providing excellent thermal isolation of the YBCO thermometer. The responsivity over a 0.2 - 2.9 THz bandwidth was 2900 V/W, the *NEP* was $9 \cdot 10^{-12}$ WHz$^{-0.5}$ and the time constant was below 10 µs. Recently, Hammar *et al.* [66] presented a broadband direct YBCO microbolometer. At 77 K they measured a responsivity of 190 V/W for a 1.5µm×1.5µm device from 330 GHz - 1.63 THz. They calculated the response time to 300 ps. An optical noise equivalent power of 20 pWHz$^{-0.5}$ for modulation frequencies between 500 Hz and 100 kHz was measured.

A detailed overview of the characteristics of superconducting YBCO high-T_c microbolometers is given in Table 2.3 revealing that minimum response times at THz wavelengths reported up to date are 300 ps.

Summary

Microbolometers embedded in planar THz antennas are the most promising bolometer-type technology to realize sensitive as well as fast detectors for picosecond time-domain measurements for the THz frequency range. The responsivity values are very high reaching 10^9 V/W for low-T_c and 2900 V/W for high-T_c microbolometers. The shortest response time reported at THz wavelengths is 165 ps for a NbN microbolometer operated at 4.2 K [58]. For YBCO the minimum reported response time of 300 ps was calculated for a YBCO microbolometer operated at 77 K [65].

A detailed experimental study concerning the YBCO response time in the THz frequency range was still missing and was performed in this work (chapter 6 and 7).

Beside the bolometer-technology other detector-types are employed successfully as direct detectors in the THz frequency range which are discussed in the following sections.

2.2. Other detector concepts - Competing technologies

The widespread bolometer detectors were already extensively studied as direct detectors in the THz frequency range as discussed in section 2.1. However, also non-bolometric detectors were successfully employed in direct detector applications in the past which are discussed in this section. The most famous representative is the Schottky-barrier diode detector, but also Golay cells, pyroelectric detectors and photoconductive switches are used as direct THz detectors and are evaluated in this section with the emphasis on fast time-domain measurements over a broad THz frequency range with a wide dynamic range.

2.2.1. Schottky-barrier diodes

Metal-semiconductor junctions with Schottky barriers are basic elements in THz technologies [67]. They are used as direct detectors and in particular very successful as nonlinear elements in heterodyne receiver mixers at temperatures between 4 and 300 K [68, 69].

Historically, first Schottky-barrier structures were pointed contacts of tapered metal wires with a semiconductor surface [70]. In the mid of 1960s, Young and Irvin developed the first lithographically defined GaAs Schottky diodes for high frequency applications [71]. The basic whiskered diode structure greatly improved the quality of the diode due to the inherently low capacity of the whisker contact [70].

The cross-section of a whisker-contacted Schottky-barrier diode (SBD) with the equivalent circuit is shown in Fig. 2.2 [72]. It consists of a junction between a platinum anode and n-GaAs epitaxial layer. Detection occurs in the nonlinear junction resistance R_j. The diode series resistance R_s and the voltage-dependent junction capacitance C_j are parasitic elements which degrade the performance.

For SBD detector operation, the series resistance and shunt capacitance influence the cut-off frequency $f_{co} = (2\pi R_s C_j)^{-1}$ which should be notably higher compared to the operation frequency f. The diode series resistance R_s is often used as a figure of merit for SBDs, but it does not represent a suitable parameter at THz frequencies. At lower frequency range ($f < 0.1$ THz), the operation of Schottky barrier diodes is well understood and can be described by mixer theory taking into account Schottky diode stray parameters (diode variable capacitance, diode series resistance). However, in the sub-mm (THz) range the design and performance of the devices become increasingly complex. At higher frequencies there appear several parasitic mechanisms credit not only, e.g., with skin effect, but also with high-frequency processes in semiconductor material such as

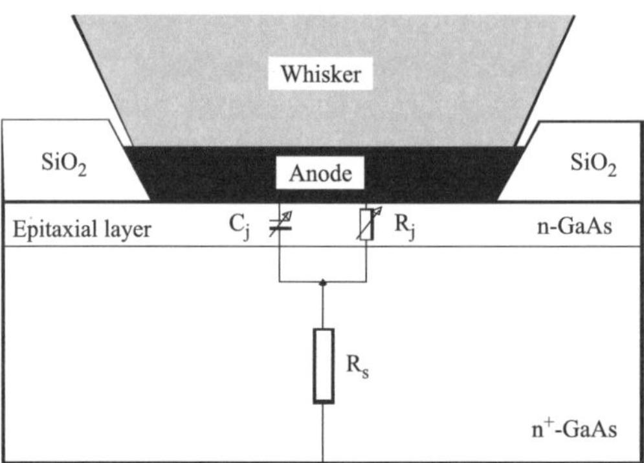

Fig. 2.2.: Crosssection of a GaAs Schottky barrier whisker contacted diode with equivalent circuit of the junction [72].

carrier scattering, carrier transit time through the barrier, dielectric relaxation, etc., which become important, and low-frequency models should be refined [67].

The frequency dependence of the responsivity for whisker-contacted SBDs was analyzed in detail in [73]. The results of the experiment and the calculation are presented in Fig. 2.3.

The solid line shows the theoretical dependence with allowance for the skin effect, the carrier inertia, the plasma resonance in the epitaxial layer (f_{pe}) and in the substrate (f_{ps}), the phonon absorption (f_t and f_l are the frequencies of the transverse and longitudinal polar optical phonons), and the transit effects. The dashed line shows the same without allowance for the transit effect. The experimental results are related to SBDs with different anode shapes (shown in Fig. 2.3). A satisfactory agreement between the experiment and the calculation takes place. It can be added that, owing to the further improvement of the antenna, the videodetector sensitivity presented in Fig. 2.3 was increased by one order of magnitude near 1 THz (up to approximately 350 V/W) [73]. The noise amounted to 20 - 100 $nVHz^{-0.5}$ at a frequency of 100 kHz, and the *NEP* values were in the range of $\approx 3 \cdot 10^{-10}$ - 10^{-8} $WHz^{-0.5}$ at a frequency of 891 GHz [73]. A further improvement in sensitivity of a GaAs SBD was obtained in [74]. The maximum value of the responsivity is 2000 V/W at a frequency of 1.4 THz and 60 V/W at a frequency of 2.54 THz. However, for the SBDs discussed so far the coupling of the THz radiation was realized by the whisker contact as dipole antenna resulting in a narrow radiation bandwidth of the SBD.

Due to limitations of the whisker technology, such as constraints on design and repeatability, starting in the 1980s efforts were made to produce planar Schottky diodes [70]. Nowadays, the whisker diodes are almost completely replaced by planar diodes [70]. Using advanced technology elaborated recently, the diodes are integrated with many passive circuit elements (impedance matching, filters and waveguide probes) onto the same substrate [75]. By improving the mechanical arrangement and reducing loss, the planar technology is pushed well beyond 300 GHz up to several THz [70]. To achieve good performances at high frequencies, the diode area should be small. Reducing junction area one reduces junction capacitances to increase operation frequency. But at the same

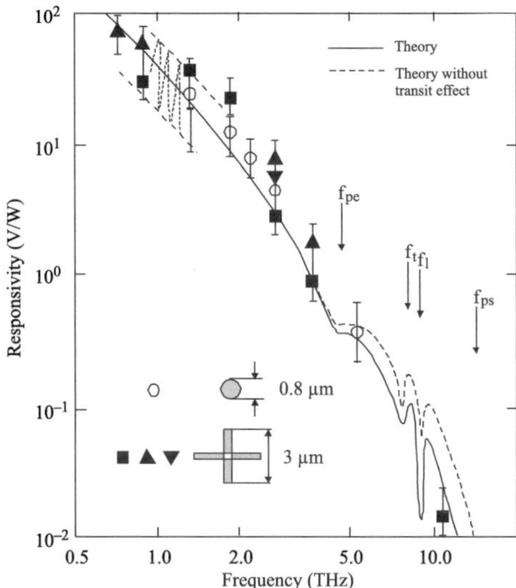

Fig. 2.3.: Dependence of the SBDs voltage sensitivity on the radiation frequency for different anode shapes according to [73].

time one increases the series resistance. The state-of-the-art devices have anode diameters about 0.25 µm and capacitances of 0.25 fF [70].

Nowadays, two different concepts are mainly used for the integration of the SBD: a waveguide-based integration or a quasi-optical concept with broadband THz antenna and silicon lens. Waveguide-based Schottky diode detectors have typical responsivities ranging from 1000 - 4000 V/W over the frequency range of 100 GHz to 1 THz [76]. However, the frequency bandwidth of waveguide-based detectors is limited by fundamental mode waveguide operation to approximately 50% fractional bandwidth. If a wider operation bandwidth is desirable, a quasi-optical detector concept is required at the cost of a reduced sensitivity. In InP-based SBDs with wide-band log-periodic toothed antenna, responsivity values of 1000 V/W at 0.3 THz and 125 V/W at 1.2 THz were reached [77]. The integration of a zero-bias SBD from Virginia Diodes, Inc. in a sinuous antenna showed responsivity values of 300 - 1000 V/W covering the frequency range of 150 - 440 GHz [76]. Semenov *et al.* [78] embedded a zero-bias SBD made from InGaAs/InP in a planar log-spiral THz antenna. Successful measurements at the synchrotron ANKA at KIT were performed resolving single pulses with a FWHM of 200 ps limited by the readout electronics. The response of the detector was nonlinear. The sensitivity decreased from $\approx$ 200 V/W at low input powers to 120 V/W for input powers larger than 20 µW. Taking into account optical losses an intrinsic diode sensitivity of larger than 10^3 V/W was estimated. The upper limit for the *NEP* of the Schottky diode was estimated to 10^{-11} WHz$^{-0.5}$ [78].

Commercially available Schottky diodes based on waveguide-coupled diodes provide rather high responsivity values in particular for low radiation frequencies (3000 V/W at 100 GHz, 100 V/W at 1 THz) but with a small operation bandwidth (about 100 GHz). The quasi-optical detectors cover a frequency range from 0.1 - 1 THz with a moderate responsivity value of 500 V/W [79]. The time resolution is below the nanosecond time scale. The dynamic range amounts to 25 dB [79].

In summary, SBDs are somewhat less sensitive than superconducting detectors, but they do not require cooling leading to cheap and handy detectors. Compared to room-temperature terahertz detectors such as Golay cells, Schottky-diode detectors have much shorter response times in the picosecond range. However, for a broadband radiation application their frequency-dependent responsivity, the moderate dynamic range of only 25 dB and the rather complex integration in a quasi-optical detector design are disadvantageous.

2.2.2. Golay cells

A Golay cell is an opto-acoustic room-temperature detector which operates over a very broad radiation frequency range from 0.2 - 20 THz. High responsivities of 10^4 V/W at a broad dynamic range of 37 dB are achieved [80] which allow to use these detectors e.g. in THz imaging applications [81, 82]. However, the quite high NEP of 10^{-8} WHz$^{-0.5}$ and very slow response times in the range of 25 ms do not allow to use this sensor for sensitive and fast time-domain analysis [80].

2.2.3. Pyroelectric detectors

Pyroelectric detectors work at room-temperature in a frequency range from 0.1 - 30 THz. They offer high responsivity values in the range of 10^5 V/W. However, with NEP values of 10^{-9} WHz$^{-0.5}$ and response times in the range of 200 ms, they are not suitable for fast and sensitive measurements [83]. Pyroelectric detectors show a broad dynamic range of 40 dB [80] and are the classical counterparts of Golay cells also applied in the field of THz imaging [84].

2.2.4. Photoconductive switches

Photoconductors can provide a greater speed of response at the expense of a narrower range of operating frequencies. They operate at higher frequencies than hot electron bolometers [21]. Gallium doped Germanium (Ga:Ge) shows a photoconductive response up to wavelengths longer than any other combination of elements. In unstressed configurations, the cut-off wavelength is approximately 120 μm ($f \approx 2.5$ THz). By applying stress to the detector crystal, this can be extended to beyond 200 μm ($f \approx 1.5$ THz). A 1.2 THz signal bandwidth centered at 3 THz is typical for these devices [21].

The performance characteristics of the detector depend on the operation conditions. In low radiation background conditions minimum NEP values as small as $5 \cdot 10^{-17}$ WHz$^{-0.5}$ at $\lambda \approx 150$ μm at an operation temperature of $T = 2$ K were reported [85], though the speed of response was limited to about 20 Hz (-3 dB). In higher background laboratory applications the detector may operate at 50 kHz (-3 dB), albeit with reduced sensitivity of $8 \cdot 10^{-13}$ WHz$^{-0.5}$. The detector responsivity amounts to 300 mA/W [21].

Although this detector-technology achieves response times on the micro- and nanosecond time scale, they are not suitable for time-domain analysis on a picosecond time scale. Also the narrow range of operating frequencies is disadvantageous for the study of broadband pulses in the THz frequency range.

2.3. The new YBa$_2$Cu$_3$O$_{7-x}$ direct THz detector technology

Recent developments in the area of pulsed THz sources (section 1 and 7.2) demand ultra-fast temporal resolution of direct THz detectors on a picosecond time scale. Further characteristics of pulsed THz sources, like e.g. electron storage rings, are a broadband emission from 0.2 - 2 THz and a very broad emitting power range (section 1 and 7.2). Therefore, a detector technology embedded in a broadband THz antenna is required which shows high sensitivities and responsivities for small emitting powers as well as a broad dynamic detection range.

Summary of existing detector technologies

As demonstrated in this chapter a variety of direct THz detector technologies already exists which were mainly optimized for high sensitivities achieving *NEP* values well below 10^{-16} WHz$^{-0.5}$ (see section 2.1.2). Detector technologies currently used at electron storage rings in the THz frequency range are either hot-electron bolometers [58, 86] or Schottky diodes [78].

Commercially available InSb hot-electron bolometers are operating at 4.2 K and are very sensitive reaching *NEP* values of $5 \cdot 10^{-13}$ WHz$^{-0.5}$ (section 2.1.3 and [21]). However, with a minimum response time on the microsecond time scale these devices are too slow for direct picosecond pulse length observation. Superconducting NbN hot-electron bolometers, also operating at 4.2 K, show much shorter response times in the range of 50 ps [86, 87] with comparable *NEP* values of 10^{-12} WHz$^{-0.5}$ [88]. However, even these fastest hot-electron bolometers are too slow for the analysis of the temporal evolution of the picosecond coherent synchrotron radiation pulses.

Room-temperature Schottky diodes have intrinsic time constants of single picoseconds fulfilling the requirement of ultimate time resolution. However, to couple broadband radiation to a Schottky diode it has to be integrated in a quasi-optical detector design with a broadband THz antenna which is rather complex and leads to reduced responsivity values (section 2.2.1). Nevertheless, with *NEP* values of 10^{-11}-10^{-12} WHz$^{-0.5}$ at room-temperature and moderate responsivity values of 500 V/W [79] they are still sensitive enough to detect the brilliant THz coherent synchrotron radiation pulses. However, the moderate dynamic range of ≈ 25 dB [79] is a drawback of this technology.

Fig. 2.4.: Multilayer structure and final layout of the new YBCO THz direct detector technology.

Characteristics of the new YBa$_2$Cu$_3$O$_{7-x}$ detector technology

Thus, the goal of this work, was the development of a new detector technology which allows for *ultra-fast time-domain measurements on a picosecond time scale*. At the same time, the integration in a broadband THz antenna is required to encompass a *broad frequency detection range*. Furthermore, the detector should show *high responsivity and sensitivity* values together with a *broad detection dynamic range*. Last but not least, a detector technology *not requiring liquid helium* for cooling significantly reduces complexity and costs of the final detection system.

A promising candidate for this task is the high-temperature superconductor YBa$_2$Cu$_3$O$_{7-x}$ (YBCO). In 1994, Richards [18] mentioned already, that fast high-T_c YBCO detectors are potentially useful with pulsed infrared sources such as synchrotron storage rings. The ultra-fast electron relaxation time of this material (section 3.2.1) allows for measurements on a picosecond time scale. Furthermore, the integration of the YBCO detecting element in a THz antenna is straightforward. In terms of sensitivity and responsivity promising results were already achieved in the past as demonstrated in section 2.1.4. With critical temperatures of YBCO well above 77 K, low-cost liquid nitrogen or cryogen-free systems can be used for cooling.

YBa$_2$Cu$_3$O$_{7-x}$ detector layout

These arguments motivate further studies on the development of YBCO thin-film THz detectors for picosecond time-domain measurements which were carried out and are presented in this work. The idea is to grow ultra-thin YBCO films between 10 and 100 nm on a substrate with low dielectric losses at THz frequencies due to the rearside-illumination of the final detector. In this work, sapphire substrate is chosen which requires the use of buffer layers to reduce the crystalline mismatch between the YBCO thin film and the substrate. For protection of the superconducting detector element during the patterning process a passivation layer is used. Finally, the THz antenna is patterned from a gold (Au) thin film which results in the multilayer structure and the final YBCO THz detector layout as shown in Fig. 2.4. The deposition, characterization and optimization of the YBCO thin films is discussed in chapter 4. Details on the patterning technology for the micrometer and sub-micrometer sized YBCO detecting elements embedded in the Au THz antenna and their DC characterisation are given in chapter 5.

Detection mechanism in $YBa_2Cu_3O_{7-x}$ detectors

The THz radiation is captured by the planar antenna and transferred into a radio frequency (rf)-current flowing through the YBCO detecting element. To optimize the detector's responsivity, the temporal resolution as well as the detection dynamic range, it is crucial to understand the detection mechanism caused by the rf-current in the detecting element. One key parameter influencing on the detection mechanism is the absorbed photon energy level which may be below or above the superconducting energy gap of the detector material. Different response mechanism for over-gap excitations in YBCO, i.e. for high photon energies down to the infrared frequency range, were already studied in detail and are discussed in section 3.2. For sub-gap excitations, such as photons in the THz frequency range, only few reports exist up to now, which are summarized in section 3.3. The experimental study for over-gap and sub-gap excitations of YBCO detectors developed in this work is then reported in chapter 6 before several successful applications of our new YBCO detector technology are discussed in chapter 7.

3. Detection mechanism in YBCO thin-films from optical to THz wavelengths

To develop and optimize a radiation detector technology, a clear understanding of the dynamic processes responsible for the radiation detection is required. This chapter gives an overview of different detection mechanism reported up to date in YBCO thin-film microbridges. The radiation frequency range from optical to infrared radiation is the most-studied and thus best-understood where YBCO microbridges are applied as bolometer detectors. After a general introduction of the YBCO high-temperature superconductor and its characteristics important for radiation detection, the existing models valid in the optical and infrared frequency range for over-gap excitations will be discussed. Furthermore, first existing reports of the extension of the studies to the THz frequency range will be presented revealing first indications for a detection mechanism based on magnetic vortices at these wavelengths.

3.1. The high-temperature superconductor $YBa_2Cu_3O_{7-x}$

In April 1986, Bednorz and Müller discovered the high-temperature superconductivity measuring a critical temperature of 35 K of Lanthanum-Barium-Copperoxide ($La_{1.85}Ba_{0.15}CuO_4$) [89]. Soon after that, several other compounds were discovered which showed critical temperatures above the boiling temperature of liquid nitrogen (LN_2). The 123-system $RBa_2Cu_3O_{7-x}$, in which R represents nearly all rare earths, is the most common one called cuprates. The most common representative of this system is $YBa_2Cu_3O_{7-x}$ (YBCO) which was discovered in 1987 by Wu *et al.* [90] and was used in this work.

The perovskite-like crystalline lattice of YBCO has a characteristic layered structure (see Fig. 3.1). The two copper oxide planes which are separated by an yttrium atom are characteristic for the YBCO unit cell followed by the barium oxide planes. The unit cell is enclosed on both sides by the *b*-axis-oriented copper oxide chains [92]. Above 700°C YBCO has an orthorhombic crystalline lattice and is twinned in the *ab*-plane. With reduced oxygen content more and more oxygen vacancies appear in the chains. The critical temperature decreases with decreasing oxygen content after an initial increase and a maximum of the critical temperature at a level of $x \approx 0.1$ [92]. Above $x \approx 0.6$ YBCO is no longer superconducting. In this range of oxygen content the material is an antiferromagnetic insulator [93]. Due to the increasing oxygen vacancies in the chains the orthorhombic crystalline lattice transfers to a tetragonal structure [92].

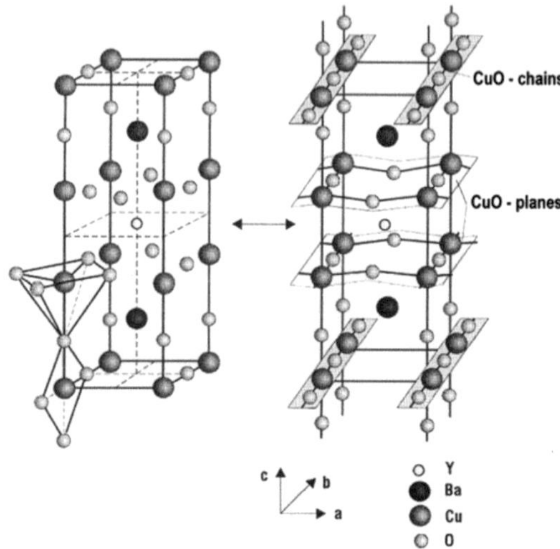

Fig. 3.1.: Schematic of the YBCO crystalline lattice. Left side shows the unit cell as well as the link between the Cu- and O-atoms. On the right side the structure with the CuO_2-planes, BaO-planes and CuO-chains is emphasized. The lattice constants at $x = 0.15$ are $a = 0.38282$ nm, $b = 0.38897$ nm and $c = 1.16944$ nm [91].

The lattice constants for $YBa_2Cu_3O_{6.85}$ with $x = 0.15$ are [91]:

a-axis: 0.38282 nm

b-axis: 0.38897 nm

c-axis: 1.16944 nm.

Since the discovery of YBCO many studies were carried out to understand the mechanism responsible for the high critical temperature of 92 K. One approach was to substitute yttrium with other rare earths [92]. However, no significant influence was found, showing that the coupling between the CuO_2-planes might only have a minor influence on the superconductivity. Up to date, there exists no comprehensive theory describing superconductivity in YBCO. However, detailed studies were carried out to analyze the intrinsic material characteristics of the high-temperature superconductor YBCO which are presented in the following chapters.

3.1.1. Magnetic properties

The high-temperature superconductor YBCO is a type-II superconductor and as such, its phenomenology is dominated by the presence of magnetic flux lines, so-called vortices, over most of the phase-diagram. The mean-field version of the latter was constructed by Abrikosov and is shown in Fig. 3.2 [94]. This mean-field H-T phase diagram comprises a Meissner phase characterized by complete flux expulsion at low magnetic fields $H < H_{c1}$, separated from the mixed (or Shubnikov) phase at higher fields $H > H_{c1}$, where the magnetic field penetrates the superconductor in the form of vortices [94]. The magnetic flux enclosed in a vortex is quantized in units of

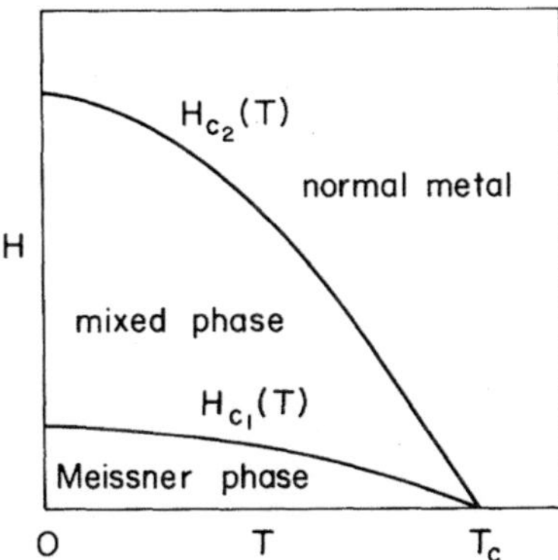

Fig. 3.2.: Mean-field phase diagram of a type-II superconductor comprising a normal-metallic phase at high fields and temperatures, separated by the upper critical-field line $H_{c2}(T)$ from the mixed or Shubnikov phase, which in turn is separated by the lower critical-field line $H_{c1}(T)$ from the Meissner-Ochsenfeld phase at low temperatures and fields [94].

the flux quantum $\Phi_0 = h/2e \approx 2.07 \cdot 10^{-15}$ Vs [94] where h is the Planck's constant and e is the elementary charge. With increasing field the density of flux lines increases until the vortex cores overlap when the upper critical field H_{c2} is reached.

Furthermore, dynamic properties of the vortex system have to be taken into account. When an external current density is applied to the vortex system, the flux lines start to move under the action of the Lorentz force [94]. This movement causes dissipation due to the appearance of a finite electric field as a consequence of the flux motion. Details can be found in [94]. The analysis of the magnetic properties of our YBCO thin films is presented in section 4.3, whereas the consequences for the detection mechanism in the developed YBCO detectors are discussed in section 6.4.

3.1.2. Optical properties

Broadband detection in the optical frequency range with high-T_c superconductors is anticipated because of their low reflectivity (less than 50%) from ultraviolet to infrared [95]. The reflectivity data of YBCO is displayed in Fig. 3.3(a). For photon energies between 1 and 2.5 eV the reflectivity is below 10%. At a photon energy of 100 meV (corresponding to 24 THz) 50% of the incoming photons are already reflected. Thus, in the THz frequency range, an absorbing layer or an antenna is required to couple THz radiation to the YBCO detector, as discussed in section 2.1.2 and 2.1.4, respectively. The spectral transmittance of c-axis oriented films is shown in Fig. 3.3(b).

3.1.3. Thermal properties

Knowledge of the thermal conductivity and specific heat capacity of YBCO is required to design a superconducting detector properly. As discussed in equation 2.7 the ratio of the specific heat capacity and the thermal conductivity to the heat sink determines the relaxation time of a bolometer.

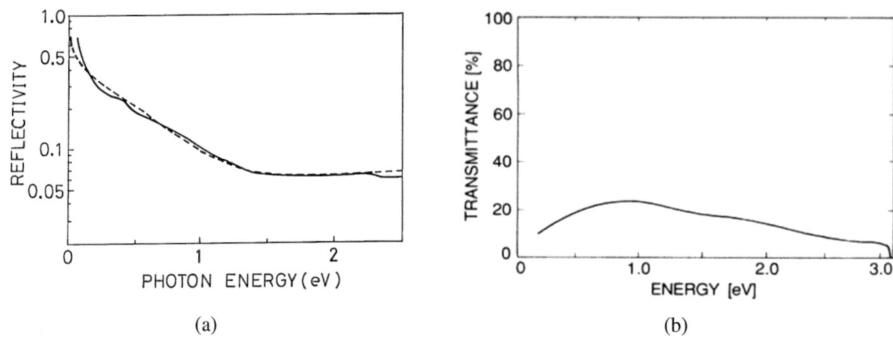

Fig. 3.3.: (a) Reflectivity spectrum of polycrystalline $YBa_2Cu_3O_7$ at room temperature [96]. (b) Infrared to ultraviolet transmittance spectrum of 180 nm thick, c-axis-oriented YBCO film [97].

The temperature dependence of the thermal conductivity of YBCO bulk material is illustrated in Fig. 3.4(a). It is important to mention that the thermal conductivity to the heat sink, responsible for the relaxation of a bolometer, is probably mainly influenced by the film interfaces and less by the bulk conductivity of the material itself.

Fig. 3.4(b) shows the specific heat capacity near the superconducting transition. With decreasing temperature the YBCO specific heat decreases leading to an increase of the bolometer's responsivity (see equation 2.6).

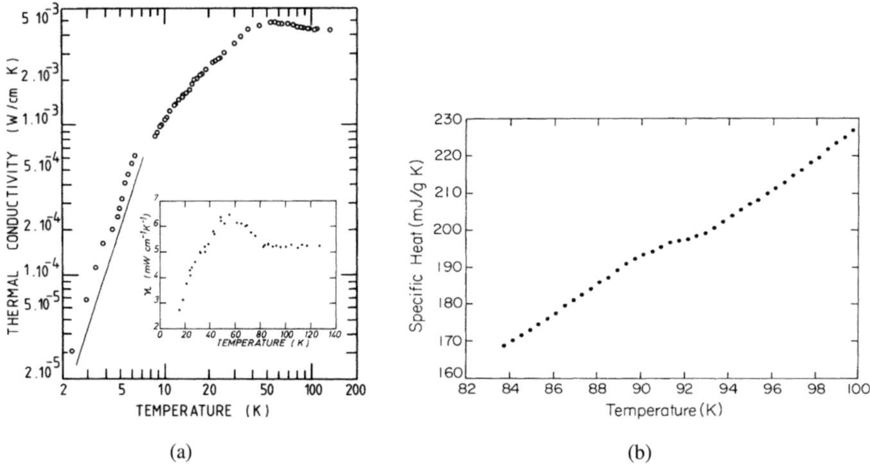

Fig. 3.4.: (a) Thermal conductivity against temperature down to 2 K for a YBCO bulk sample. The bold line indicates the T^3 dependence. The inset shows the data around T_c on a linear scale [98]. (b) AC specific heat data near the superconducting transition of YBCO [99].

3.1.4. Superconducting energy gap

A large variety of studies has been already performed to determine the superconducting energy gap of YBCO. Some groups estimated the gap by fitting their optical data to the Mattis-Bardeen theory. E.g. Bonn *et al.* [91] interpret their optical data to show an energy gap of 25 meV. Thomas *et al.* [100] found an energy gap of 27 meV. Another method to determine the superconducting energy gap is from tunneling measurements. Kirtley *et al.* [101] found 32 meV and Crommie *et al.* [102] found 30 meV. Fogelstroem *et al.* [103] determined the gap value to be between 17 and 25 meV. The data shows some scatter ranging from 17 up to 30 meV. However, according to [104], a range between 20 and 25 meV for bulk YBCO should be a safe estimate.

3.2. Excitations above the energy gap - Optical wavelengths

Due to the lack of available brilliant, picosecond pulsed sources in the THz frequency range, the study of the response mechanism in YBCO thin film detectors was limited to optical and infrared wavelengths in the past. For these wavelengths, i.e. for over-gap excitations of the YBCO detector, different theories were developed and are discussed in this section. Beside the well-known hot-electron effect (section 3.2.1) also response mechanism including the dynamics of vortices (section 3.2.2 and 3.2.3) or the kinetic inductance of the YBCO detectors (section 3.2.4) are reviewed.

3.2.1. Hot-electron effect - Two-temperature model

The photoresponse of superconducting YBCO thin films was already extensively studied in the spectral range from optical to infrared wavelengths [9–12]. Typically, films were patterned to form micrometer-wide bridges with thicknesses between 40 and 100 nm. In all these experiments the photon energy of incoming radiation was above the superconducting energy gap in YBCO, and hence the radiation was directly absorbed by electrons. The photoresponse in this wavelength range is based on the hot-electron effect and is well described by the two-temperature model [87]. The main steps of the hot-electron phenomenon after absorption of optical pulsed radiation are depicted in Fig. 3.5. The general idea is that electrons excited by absorbed radiation destroy Cooper pairs and produce hot electrons in the microbridge. An increased quasi-equilibrium electron temperature T_e is established within the thermalization time τ_T. The excess energy is then transferred to the phonons within the electron-phonon interaction time τ_{ep} that increases the quasi-equilibrium phonon temperature T_p (hence 2T model). The hot-electron model is valid for non-equilibrium superconductors maintained at a temperature T near the superconducting transition temperature T_c, where quasiparticles and phonons can be described by thermal, normal-state distribution functions, each with its own effective temperature. The electron and phonon effective temperatures (T_e and T_p) are assumed to be established instantly and uniformly throughout the whole specimen.

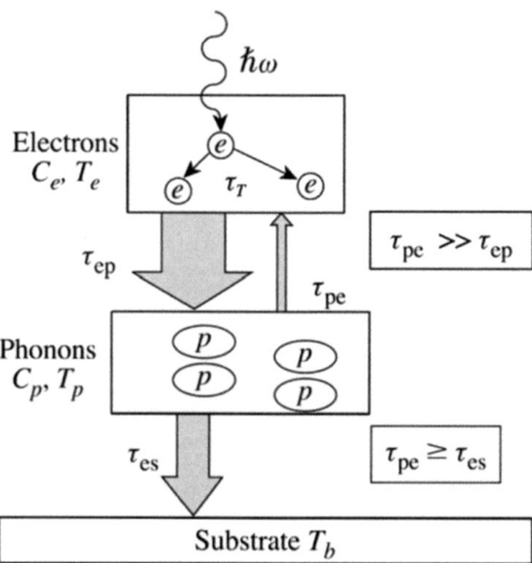

Fig. 3.5.: Thermalization scheme showing subsequent channels of the energy transfer in a hot-electron device that relaxes towards global equilibrium [87].

This assumption implies that a rapid thermalization mechanism exists inside the electron and phonon system.

Introducing characteristic times of the energy exchange between the sub-systems reduces the problem of the global equilibrium recovery to a pair of coupled heat-balance equations for T_e and T_p. The intrinsic thermalization time τ_T should be short compared to energy exchange times. This two-temperature (2T) approach is only valid in the vicinity of T_c. Below T_c, the electron specific heat exhibits an exponential temperature dependence that makes equations nonlinear for even small deviations from equilibrium. Near T_c, however, the description can be simplified. At this temperature, the superconducting energy gap is suppressed, concentration of Cooper pairs is very small and unpaired electrons exhibit no significant superconducting peculiarities: they are regarded as normal electrons having the ordinary Fermi distribution function. In the normal state, the specific heat of the electrons has a much weaker temperature dependence, which can be neglected for small deviations of T_c from equilibrium. With these assumptions, the equations describing the hot-electron effect in superconductors become linear and can be written according to [87] as

$$\frac{dT_e}{dt} = -\frac{T_e - T_p}{\tau_{ep}} + \frac{1}{C_e} W(t) \tag{3.1}$$

$$\frac{dT_p}{dt} = -\frac{C_e}{C_p}\frac{T_e - T_p}{\tau_{ep}} - \frac{T_p - T_b}{\tau_{es}} \tag{3.2}$$

where $W(T)$ represents the external perturbation (i.e. the power per unit volume absorbed by the electron sub-system), τ_{ep} and τ_{es} are the electron energy relaxation time via electron-phonon interaction and the time of phonon escape into the substrate, respectively. C_e and C_p are the electron and phonon specific heats, respectively and T_b is the ambient (substrate) temperature. In simple

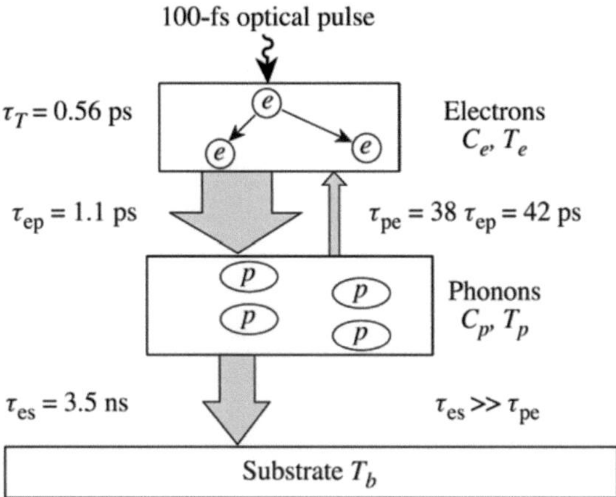

Fig. 3.6.: Hot-electron relaxation diagram [87] and characteristic time constants [106] for a 100 nm thick YBCO film.

words, equations 3.1 and 3.2 describe the rise of the electron and phonon temperature of the superconductor for a certain external perturbation as well as its relaxation back to equilibrium. To derive the 2T equations the condition of the energy-flow balance in equilibrium $\tau_{pe} = \tau_{ep}(C_p/C_e)$ was used, where τ_{pe} is the phonon-electron energy relaxation time.

An analytical solution of equations 3.1 and 3.2 was first discussed by Perrin and Vanneste [105] for sinusoidal perturbations and by Semenov *et al.* [9] for optical pulse excitations. Fig. 3.6 presents the detailed thermalization diagram for a 100 nm thick YBCO thin film excited by a 100 fs wide laser pulse. In YBCO the electron thermalization and electron relaxation times are in the sub-picosecond ($\tau_T = 0.56$ ps) and picosecond (τ_{ep}) range as it was measured by electro-optical sampling technique [106]. Within the phonon escape time τ_{es} the energy is released to the substrate, which permanently remains at the bath temperature. The phonon escape time is proportional to the film thickness on a nanosecond time scale (see Fig. 3.6).

The comparison with another superconductor typically used for hot-electron bolometers, namely NbN, reveals that the electron thermalization and relaxation is about one magnitude faster in YBCO. Il'in *et al.* [78] demonstrated measurements of the energy relaxation processes in NbN. With electro-optical sampling technique electron-phonon interaction times τ_{ep} of about 12 ps were found.

Another characteristic of YBCO is that the specific heat capacity of the phonons C_p in YBCO is much larger than the electron specific heat capacity C_e ($C_p/C_e = 38$ [106]). Consequently, as can be seen from equation 3.2, in a YBCO thin film excited on the femtosecond time scale, the non-thermal (hot-electron) and thermal, bolometric (phonon) processes are practically decoupled, with the former dominating the early stages of electron relaxation. This explains the fast picosecond component in the photoresponse of YBCO microbridges. The following slow component is largely due to the cooling of phonons via escape to the substrate.

To compare again to NbN, the ratio C_p/C_e in NbN is much smaller (C_p/C_e = 6.5 [106]) thus the energy backflow from the phonon system to the electron system can not be neglected. However, the phonon escape time to the substrate is much faster in the order of 50 ps [86, 87]. This can be explained by the ultra-thin films which are possible to fabricate from NbN in the range of single nanometers. Additionally, NbN films are in general better acoustically matched to the substrate which significantly reduces τ_{es} [87].

Bolometric response times of the order of nanoseconds in YBCO thin films have been observed by several research groups using pulsed-laser and synchrotron radiation sources from the visible to the infrared [41, 107–109]. Responses in YBCO thin films described by this detection mechanism, which includes creation of hot electrons and subsequent electron-phonon cooling, was reported for wavelengths up to $\lambda = 20$ µm ($f = 15$ THz) [41]. This study could not be extended to longer wavelengths in the past, due to the lack of brilliant pulsed sources in the THz frequency range.

3.2.2. Photo-activated flux motion

Beside the generally accepted hot-electron effect explaining the fast photoresponse on a picosecond time scale in YBCO, there have been also other theories to explain this non-bolometric component. Frenkel *et al.* studied the non-bolometric component of the pulsed optical response of laser-deposited *c*-axis oriented crystalline YBCO thin films [109, 110]. They found that below the superconducting transition the response amplitude decreased gradually (considerably slower than dR/dT), but did not go to zero which would be expected for a pure bolometric detection mechanism. This was clear evidence of a non-bolometric effect. To explain their results they suggested the theory of photoactivated flux motion [111]. Due to the strong evidence that the dissipation properties of epitaxial YBCO thin films are most likely determined by flux-creep and a flux-flow model [112, 113] Frenkel *et al.* suggested that the pulsed optical response is likely caused by the photoactivated flux creep and flux flow. Fluxons are present in the film and are moved by thermal activation [112]. The optical photons create additional flux creep and flux motion which apparently causes additional dissipation (equivalent to the increase in the resistance) and which is attributed to the non-bolometric component in the optical response [111]. The energy of photons at the wavelength of ≈ 1 µm is about 1.2 eV which is slightly larger or on the same order of magnitude as the activation energy U_0 which should be overcome to allow flux motion. U_0 for YBCO was calculated [109, 114] and measured experimentally [112, 115]. Obviously, with the temperature decrease, the activation energy increases which should decrease the amplitude of the optical response. In the transition region the response is predominantly bolometric which can be easily explained by simple lattice heating (U_0 goes to 0).

The theory of photo-activated flux motion shows that already in the past the contribution of vortices to the photoresponse was taken into account. However, the optical excitations were well above the superconducting energy gap preventing the clear separation of the hot-electron effect from a vortex-assisted photoresponse.

3.2.3. Photofluxonic model

Another approach for describing the photoresponse in superconducting thin films based on vortices was presented by Kadin *et al.* [116]. They proposed a quantum detection mechanism for photon absorption in a two-dimensional superconductor that exhibit a vortex-unbinding transition. A photon of energy $E = h\nu \gg 2\Delta$, where 2Δ is the superconducting energy gap, is absorbed by a superconducting film creating non-equilibrium quasiparticles depressing Δ thus leading to the creation of a pair of equal and opposite fluxons (or vortices), each with quantized flux $\Phi_0 = h/2e$. An applied current sweeps these fluxons to opposite edges of the film, causing a voltage pulse with time-integrated magnitude Φ_0, and leading to a time-averaged voltage responsivity given by

$$S = \frac{\Phi_0}{E} = \frac{1}{2e\nu}. \tag{3.3}$$

This two-dimensional model was applied to thin films with the thickness of much less than the Ginzburg-Landau coherence length and the magnetic penetration depth. Kadin *et al.* extended this model and argued that the three-dimensional limit of a vortex pair is a vortex ring, essentially a vortex closing on itself. A vortex ring should nucleate in much the same way as a vortex pair, with a locally depressed value of the critical current, which may be caused by the incoming photon. A Lorentz force of the current applied parallel to the axis of a vortex ring will cause this ring to expand, eventually becoming a vortex-antivortex pair. Ultimately, this will transfer a single flux quantum across the film, leading to the same responsivity S given by equation 3.3 [111].
In summary, the photofluxonic model attributes the photoresponse to dissipative movement of vortices. However, also in this approach, the excitations were assumed to be well above the superconducting energy gap.

3.2.4. Kinetic-inductance photoresponse

For superconducting detectors operated at $T \ll T_c$ Rothwarf and Taylor [117] were the first to successfully develop the phenomenological description for non-equilibrium Cooper pair recombination and breaking processes (the so-called RT model). At low temperatures, when energies of non-equilibrium quasiparticles after thermalization are spread over a narrow interval above the superconducting energy gap 2Δ, the appropriate parameters to characterize this non-equilibrium state are the number, Δn_q, of excess quasiparticles and the number, Δn_p, of excess, the so-called, 2Δ phonons. The 2Δ phonons are emitted in the Cooper pair recombination process and, since they have the energy of at least 2Δ, they are responsible for secondary breaking of Cooper pairs. Like the 2T model, the RT approach assumes that there is a quick, intrinsic thermalization mechanism inside both the quasiparticle and phonon sub-systems. When photons with energy typically much larger than 2Δ are absorbed by a superconducting film maintained at $T \ll T_c$, they produce a time-dependent population $\Delta n_q(t)$ of non-equilibrium quasiparticles, leading to a temporary decrease in the superconducting fraction of electrons, $f_{sc} = (n_0 - n_q)/n_0$, where $n_q = n_q(0) + \Delta n_q(t)$ is the

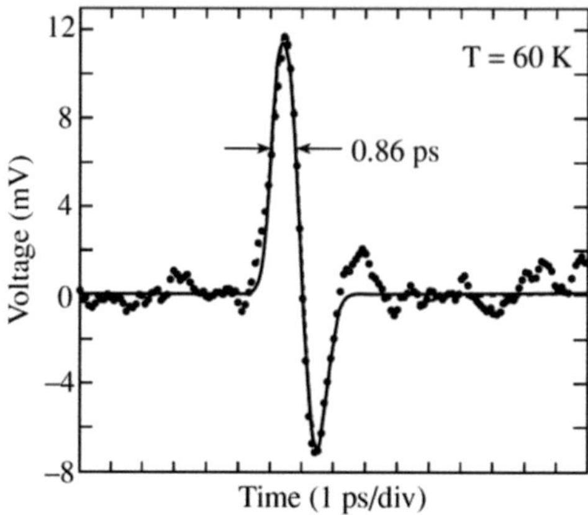

Fig. 3.7.: Kinetic-inductance response of a YBCO detector to 100 fs wide optical pulses, operated at low temperatures (60 K) in the superconducting state. Dots correspond to the experimental data, while simulated transients were obtained using the RT model (solid line) [106].

instant concentration of the quasiparticles, $n_q(0)$ is their equilibrium concentration and n_0 is the total concentration of electrons. Because the pairs are characterized by nonzero inertia, this process can be modelled as time-varying kinetic inductance [118, 119]:

$$L_{kin}(t) = \frac{L_{kin}(0)}{f_{sc}},$$
(3.4)

where $L_{kin}(0) = \mu_0 \lambda_L^2/d$ is the equilibrium value per unit area of the film, λ_L is the magnetic penetration depth and d is the film thickness. The L_{kin} of a superconducting film makes it possible to monitor the concentration of Cooper pairs. In a current-biased superconducting film, after the destruction of a certain number of Cooper pairs, the remaining pairs accelerate to carry the same bias current. Because of nonzero inertia of pairs, acceleration requires an electric field. This intrinsically generated electric field is seen from the exterior as a voltage pulse developing across the film. Mathematically, this voltage transient is given by

$$V_{kin} = I\frac{dL_{kin}}{dt}.$$
(3.5)

A typical kinetic-inductance response for a YBCO thin-film is demonstrated in Fig. 3.7 [106]. The positive part of the transient represents the Cooper pair breaking process, while the negative part represents pair recombination.

The theory of kinetic-inductance photoresponse takes into account the change of the Cooper pair density induced by the absorption of over-gap photons. To observe this phenomenon the detector operation temperature has to be well below its critical temperature.

3.3. Excitations below the energy gap - Vortex dynamics at THz wavelengths

Although slight differences in the response of a YBCO microbridge to optical and infrared radiation pulses [120] as well as nonbolometric features of the response to THz radiation [121] have been noted in earlier publications, it was generally accepted that the detection mechanism remains unchanged even for sub-millimeter wavelengths (terahertz frequency range). The THz radiation pulses available at that time were not sufficiently short to allow for quantitative evaluation of the non-bolometric features and to find out to what extent they may affect the response time of the microbridge. First the discovery of intensive coherent synchrotron radiation in the form of picosecond pulses has made it possible to analyze the response time of a YBCO microbridge to sub-gap excitation which was carried out in this work.

As discussed in section 3.1.4 the energy gap of optimally doped single-crystalline YBCO is 20 - 25 meV at zero temperature. For typical optical wavelengths of $\lambda = 800$ nm the photon energy of 1.55 eV is well above the energy gap and suffices for destroying Cooper pairs. To the contrary, in the THz frequency range the photon energy is smaller than the energy gap. The major energy of coherent synchrotron radiation generated by 1 ps long pulses is emitted in the frequency range below 1 THz which corresponds to photon energies below 4.1 meV. Even if we take into account possible contributions of the d-wave pairing and the second energy gap which leads to a somewhat smaller effective energy gap for temperatures close to the superconducting transition, the THz photon energies are below the energy gap for most operation conditions. Thus, absorption of the THz photon by a Cooper pair with subsequent breaking of the latter can be safely neglected.

In the past, already several theories have been taken vortex dynamics into account responsible for the photoresponse in high-T_c YBCO (see section 3.2). However, in all these scenarios the response was excited by photons well above the superconductor energy gap.

A detailed study for pulsed excitations above and below the energy gap for YBCO thin films was missing prevailing conclusive theories for pulsed sub-gap excitations in the THz frequency range. In this work, a detailed experimental study for optical and THz pulsed excitations is presented revealing clear differences for these two regimes (see chapter 6).

3.4. Conclusions

In this chapter, existing models for the detection mechanism in the high-temperature superconductor YBCO were summarized. It was emphasized that up to now, the majority of these studies were concentrated on photon excitations above the superconducting energy gap of YBCO (section 3.2). The most common theory is the 2T model which describes the photon absorption by Cooper pair breaking and electron heating with subsequent electron and phonon relaxation processes. However, also different models were discussed taking into account the typical type-II superconductor characteristic of dynamic vortex motion. Nevertheless, for all these studies an over-gap excitation of YBCO was presumed leading to photon absorption in the Cooper pair and quasiparticle sub-system of the superconductor.

The very few works discussing sub-energy gap excitations were summarized in section 3.3. First indications of a non-bolometric detection mechanism were already found there. However, the lack of a brilliant, pulsed THz source prevailed a detailed study of the detection mechanism of YBCO thin-film detectors to sub-gap excitations which was carried out and will be discussed in this work.

4. Fabrication technology

For the development of high-quality superconducting detectors the first step is the optimization of the superconducting thin-film deposition. For fast and sensitive YBCO detectors ultra-thin films with thicknesses between 10 and 100 nm are required.

In this work, the pulsed-laser deposition (PLD) technique was used to deposit these superconducting thin films. In the following chapter, the PLD setup will be presented and the optimization of the YBCO thin-film deposition will be discussed in detail. The deposited YBCO thin films are characterized by measurements of the superconducting characteristics, also in magnetic field.

4.1. YBCO deposition process - Pulsed-Laser Deposition

There are many processes for depositing YBCO superconducting films. Typical ones are PLD, thermal evaporation, rf-sputtering, metalorganic chemical vapour deposition, liquid-phase epitaxy, *ex situ* electron-beam processing, metalorganic deposition and the sol-gel process. Among them, PLD is predominant for fabricating thin and ultra-thin YBCO films [122–124].

PLD has been studied since the development of high-power lasers about 40 years ago [125]. Excimer lasers are currently the lasers of choice due to their ultraviolet emission. With regard to PLD, UV light provides two factors unavailable to the infrared light of CO_2 lasers. The first is the ability to efficiently breakup the irradiated material into atoms, ions and molecules. Such species are often necessary for the production of high-quality deposited films of the correct crystalline phase and stoichiometry. Secondly is the ability to transmit plasmas if and when they form. The free electrons in a plasma will strongly absorb laser radiation thus preventing it from reaching the surface. The absorption occurs by a process known as inverse Bremsstrahlung, with the absorption intensity being proportional to the square of the laser wavelength. For the same plasma density, IR light is highly attenuated whereas ultraviolet excimer radiation will pass with little absorption. This permits much greater efficiency for the ablation process [126].

4.1.1. The Pulsed-Laser Deposition setup

The YBCO thin films discussed in this work have been fabricated using the PLD technique. An existing setup [127, 128] was recommissioned and optimized for the fabrication of ultra thin films. The setup is divided in three parts: the excimer laser with optical components to guide the laser light in the deposition chamber, the deposition chamber with target-carousel and oxygen-compatible heater and an electronic rack with PC to control the deposition process. A picture of the whole PLD setup is shown in Fig. 4.1.

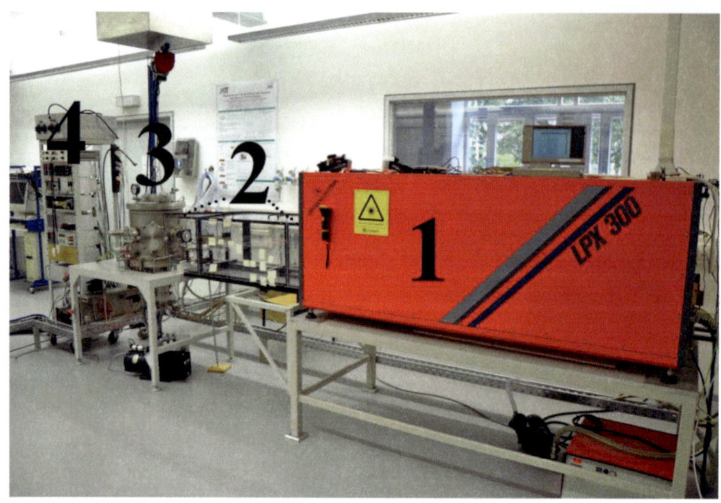

Fig. 4.1.: PLD setup at IMS: The UV light is emitted by the excimer laser (1), guided via mirrors to a lens (2) which focuses the light onto the target in the vacuum chamber (3). The electronics (4) are used to control the deposition process.

Optical components

The radiation source is a KrF excimer laser of Lambda Physics (model LPX 305iF) with a wavelength of $\lambda = 248$ nm. This laser has a maximum pulse energy of 1.2 J at 1 Hz repetition rate and achieves at the maximum repetition rate of 50 Hz still 1 J. The pulse duration is 25 ns. The pulse energy stability was $\pm 3\%$ after delivery in 1994 according to the data sheet. However, nowadays fluctuations in the pulse energy up to 10% appear.

Fig. 4.2 shows the intensity profile of the laser beam determined by Lambda Physics [129]. In horizontal direction the intensity is constant over 25 mm, whereas in vertical direction the intensity profile is described by a Gaussian distribution.

To make the current intensity distribution visible thermal paper was used directly after the beam exit of the laser. Fig. 4.3 shows the intensity profile of the laser beam with the different gray areas showing strong inhomogeneities within the laser spot.

Thus, directly after the laser beam exit an aperture is used to cut the less intensive parts of the laser beam profile. This is important to get a homogeneous energy distribution on the target surface.

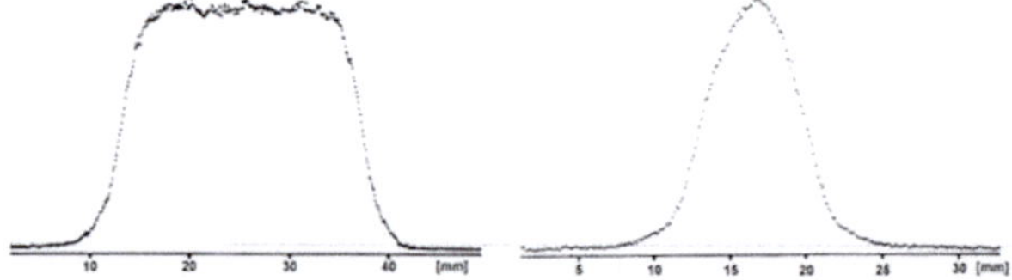

Fig. 4.2.: Intensity profile of the LPX 305iF laser in horizontal (left panel) and vertical (right panel) direction according to [129].

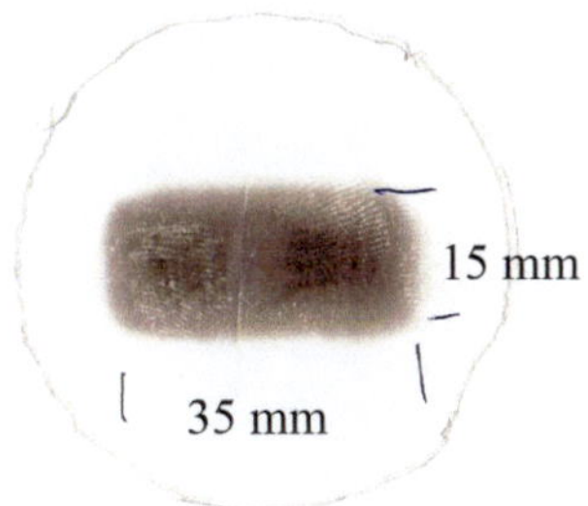

Fig. 4.3.: Intensity profile of the laser beam exit. The inhomogeneity of the intensity profile is demonstrated by the different gray areas.

The beam is guided by two mirrors, which deflect the beam by $90°$ to the window of the deposition chamber. In front of the window a lens focuses the beam onto the target. The focal length of the lens is 325 mm which results in a laser spot size on the target of 1.8×5 mm^2. The complete optical path is schematically shown in Fig. 4.4. In total, 70% of the laser energy after beam exit are lost by the optics resulting in 30% of the laser energy on the target. For the maximum laser energy of 1.2 J this results in a maximum energy density on the target of 4 J/cm^2.

Deposition chamber

A photo of the inside of the deposition chamber is shown in Fig. 4.5. The 45 cm high chamber has a diameter of 40 cm with a laser entry window (see Fig. 4.5 (1)) at 20 cm height.
The chamber is equipped with a carousel of six targets (see Fig. 4.5 (2)), each with a diameter of 1 inch. Currently four targets are installed in the chamber:

- YBCO target from Jupiter Technologies with a density of 99%,

- CeO$_2$ target from Adelwitz Technologiezentrum GmbH,

- PBCO target from Leibniz Institute for Solid State and Materials Research Dresden,

- Au target from MaTeck GmbH.

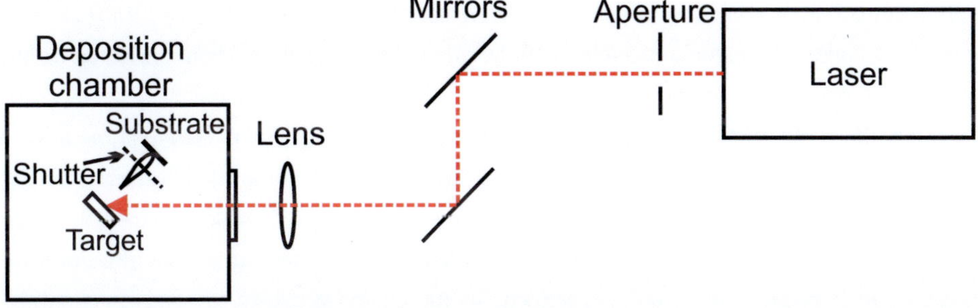

Fig. 4.4.: Scheme of the optical path of the laser beam from the laser exit to the deposition chamber.

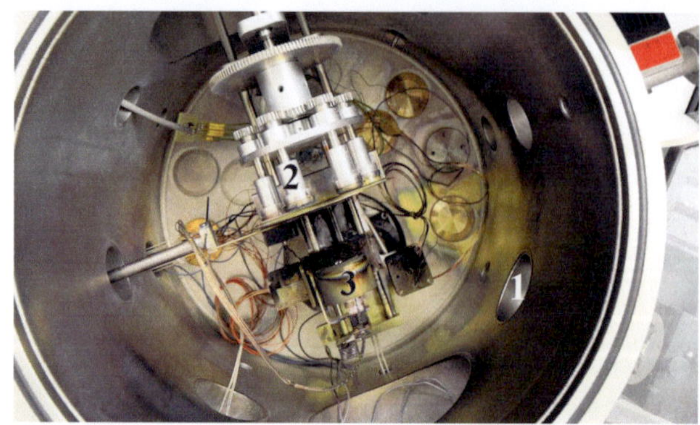

Fig. 4.5.: Inside of the deposition chamber: The laser beam enters the chamber through the window (1) and hits the target at the target carousel (2). The ablated material is deposited onto the substrates situated on the heater (3).

A DC motor rotates the targets during deposition to allow uniform ablation. The installed shutter is used for protection of the substrates during precleaning of the targets to ensure that the imperfect surface of the targets is removed right before the thin film deposition.

Since heating of the substrate to temperatures of $\approx 800°$C in oxygen atmosphere is essential for the growth of superconducting YBCO thin films, the chamber is equipped with an oxygen-compatible heater (see Fig. 4.5 (3)). The position of the heater is adjustable by four motors: the distance between the substrate and target, the height of the heater, the angle in reference to the target and the tilt from on-axis to off-axis position are freely adjustable.

A crucial improvement of the setup has been the development of an own heater fabrication techno-logy at the IMS. A photo of a ready-fabricated heater is shown in Fig. 4.6. The heater developed consists of three parts: the stainless steel heating plate (see Fig. 4.6 (1)) with an embedded heater spiral with cold ends (see Fig. 4.6 (2)), a ceramic shield (see Fig. 4.6 (3)) and an outer shielding made from stainless steel (see Fig. 4.6 (4)). The heating wire tolerates temperatures as high as $850°$C, has an outer diameter of 1.5 mm (inner diameter: 0.22 mm), a resistance of 32 Ω/m and an operation voltage of 100 V. The cold ends of the heating wire have been realized with a paste suit-able to glue metals and withstand temperatures up to $1600°$C. For the ceramic shield an ultra-high vacuum compatible ceramic has been chosen which tolerates temperatures up to $1100°$C after sin-tering. The stainless steel shield together with the other components has been fabricated in-house according to the particular recipient conditions.

A prepump filled with Fomblin oil, compatible with oxygen atmosphere, is used to pump down the chamber to pressures of $2 \cdot 10^{-2}$ mbar. In combination with a turbo pump (type Balzers TMH 260) with a pump capacity of 230 l/s a minimum base pressure of $2 \cdot 10^{-7}$ mbar is achieved. Oxygen is connected via high-vacuum lines to a flow controller which is used during the film deposition, whereas a direct connection of the oxygen line to the chamber allows to vent the chamber to 900 mbar for annealing of the samples. Nitrogen is used to vent and open the chamber e.g. after the final sample deposition.

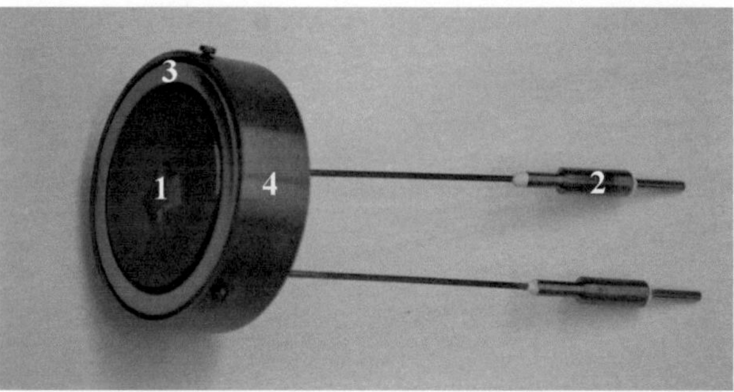

Fig. 4.6.: Oxygen-compatible heater with stainless steel heating plate (1) with embedded heater spiral with cold ends (2), a ceramic shield (3) and an outer stainless steel shielding (4).

LabVIEW controlled electronics

The PLD process is controlled by a LabVIEW program [128]. The valves, heater position, laser and pressure settings during the deposition are adjustable by the software. Furthermore, the deposition temperature as well as the annealing process are monitored by the LabVIEW program.

4.1.2. The deposition process

In spite of the small lattice mismatch of $SrTiO_3$ or $LaAlO_3$ to YBCO (1.4% and 2% respectively) [130], we choose both-side polished R-plane sapphire as substrate for our YBCO thin films with a stronger lattice mismatch of 12% [131]. This is due to the low dielectric losses of sapphire ($\varepsilon_r = 10.06$, $\tan\delta = 8.4 \cdot 10^{-6}$ at 77 K) [131] which are essential for the back illumination of our YBCO detectors at THz frequencies.

The sapphire substrate is cleaned using acetone and isopropanol in an ultrasonic bath. To ensure a good thermal contact during deposition the substrate is mounted on the heater with silver paste. After mounting, the heated substrate holder is positioned 47 mm from the target in on-axis position. The vacuum chamber is evacuated by a turbo pump to a base pressure below $1 \cdot 10^{-6}$ mbar. For the deposition the substrate is heated up to 800°C with a ramp of 700°C/hour.

Due to the crystalline mismatch between sapphire and YBCO and diffusion of aluminum into the YBCO film at high deposition temperatures of $\approx$ 850 °C [131, 132], it is not possible to grow high-quality superconducting YBCO thin films directly on sapphire. Buffer layers to improve the matching of the crystalline structure have to be used in the deposition process. The target carousel of the PLD system allows *in-situ* fabrication of multilayer samples. A typical buffer layer material for YBCO thin-film deposition on sapphire is Ceriumoxide (CeO_2). CeO_2 has a fluorite structure i.e. a cubic symmetry. Its lattice parameter ($a = 0.541$ nm) matches closely with the ab-plane of YBCO rotated by 45°. The resulting lattice mismatch is 0.2% and 1.3% along the a- and b-axis of YBCO, respectively [133].

A CeO_2 buffer layer is deposited at a substrate temperature of 800°C, an oxygen pressure of $p_{O2} = 0.9$ mbar and a distance between target and substrate of 47 mm. Pulses from the KrF

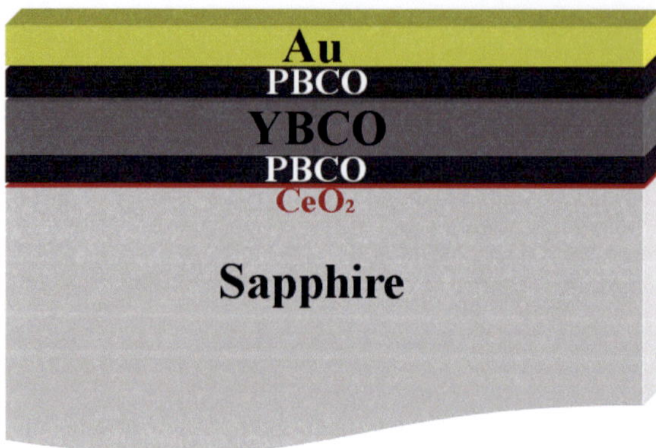

Fig. 4.7.: Schematic of the final multilayer CeO$_2$-PBCO-YBCO-PBCO-Au on sapphire substrate.

excimer laser (wavelength λ = 248 nm) are focused on the rotating CeO$_2$ target with a frequency of 10 Hz and a pulse energy density of $\approx$ 1.5 J/cm^2 resulting in a deposition rate of 0.8 nm/s.

To further improve the crystalline matching with the YBCO film, an additional buffer layer made of PrBa$_2$Cu$_3$O$_{7-\delta}$ (PBCO) is deposited on top of the CeO$_2$ layer. PBCO shows a very small lattice mismatch to YBCO ($\leq$ 1.5%) and has the same orthorhombic structure as superconducting YBCO [134]. For the deposition of the PBCO buffer layer, the deposition temperature, the target-substrate-distance and the laser energy density were kept constant, and the partial oxygen pressure was reduced to p_{O2} = 0.7 mbar. The PBCO deposition rate is 0.7 nm/s. The YBCO film is then deposited on top of the PBCO buffer layer at the same temperature, oxygen pressure and laser energy density with a target-substrate-distance of 52 mm. The YBCO deposition rate amounts to 0.8 nm/s.

For protection of the YBCO thin film a passivating PBCO layer is deposited on top of the YBCO film. After the second PBCO layer deposition the oxygen pressure is increased to 900 mbar, and the substrate temperature is ramped down to 550°C with a rate of 10°C/min. The temperature is kept constant at 550°C for 10 minutes for annealing of the obtained multilayer structure. Afterwards, the heater is ramped down to 400°C before switching off and cooling down exponentially to room temperature. The vacuum chamber is then pumped down to pressures about $5 \cdot 10^{-5}$ mbar and a Au layer is grown *in-situ* using the same PLD technique. The energy density is increased to $\approx$ 2 J/cm^2 for the Au deposition resulting in a deposition rate of 0.15 nm/s.

The final multilayer is shown in Fig. 4.7.

4.2. Optimization of YBCO thin films on sapphire substrate

Due to the good lattice matching (see section 4.1.2) high-quality YBCO thin films can be fabricated on LaAlO$_3$ and SrTiO$_3$ substrates. Zero-resistance critical temperatures between 88 and 90 K with transition widths below 1 K for 100 nm thick YBCO films were reported from different groups [10, 11, 135]. Reported critical current densities at 77 K were always above 1 MA/cm^2. Ultra-thin

Fig. 4.8.: SEM image of a YBCO-CeO$_2$ bilayer on sapphire. The typical island growth of YBCO is clearly visible.

films existing of only one layer of YBCO embedded in PBCO layers grown on SrTiO$_3$ substrate still showed a superconducting transition with a zero-resistance critical temperature of 19 K [134]. The deposition on sapphire is due to the requirement of additional buffer layers much more challenging. However, with an optimized process critical temperatures as high as 88 K with transition widths below 1 K were reported for 100 nm thick films [121]. Critical current densities of 1.5 MA/cm^2 at 77 K were reported for a CeO$_2$-YBCO bilayer on sapphire [131].

4.2.1. YBCO layer optimization

The deposition of the superconducting YBCO thin film has already been optimized in a previous work (see [127]). The deposition parameters were optimized on LaAlO$_3$ substrates and the quality of the films was evaluated amongst others using X-ray diffractometry and SEM technique. The 2θ-θ-scan revealed a very good oxygen content of our 100 nm thick films resulting in a c-axis lattice constant of 1.169 nm. The ω-scan showed a mosaic spread from 0.36° to 0.4° for the (005) peak while according to [136] a perfect crystalline lattice is achieved for 0.36°. The optimized YBCO deposition parameters in terms of laser energy density (1.5 J/cm^2), distance between target and substrate (52 mm) and partial oxygen pressure (0.7 mbar) were transferred to the deposition of YBCO on sapphire. For this, a 50 nm thick buffer layer of CeO$_2$ was introduced resulting in the successful growth of YBCO thin films on sapphire as demonstrated by the SEM picture in Fig. 4.8. The typical island growth of YBCO is visible. The detailed optimization of the multilayer deposition process required for YBCO thin-film growth on sapphire was carried out in this work and is discussed in the sections below.

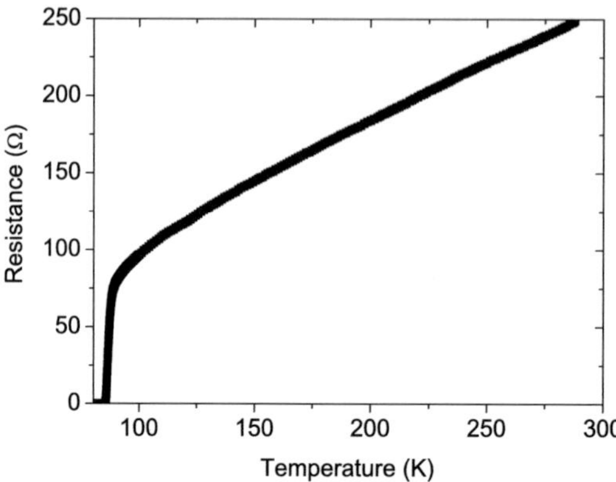

Fig. 4.9.: Dependence of resistance on temperature for a 50 nm thick YBCO thin film on sapphire with CeO_2 buffer layer.

4.2.2. CeO_2 buffer layer optimization

The first step of optimization was the determination of the optimal thickness of the CeO_2 buffer layer [PSH$^+$12b]. Whereas a thick layer allows for good crystalline growth, the longer pulsed laser deposition time leads to an increase of droplets on the sample surface resulting in a rough surface which is disadvantageous for the following layers.

A CeO_2 buffer layer with a film thickness between 8 and 50 nm was deposited on the sapphire substrate at a temperature of 800°C and an oxygen pressure of $p_{O2} = 0.9$ mbar. A YBCO thin film of 50 nm thickness was then deposited *in-situ* on top of the CeO_2 layer at the same temperature with $p_{O2} = 0.7$ mbar. After the YBCO deposition the sample was annealed and cooled down as described in section 4.1.2. The as-deposited films were characterized using a quasi four-probe measurement configuration to determine the dependence of the film resistance on temperature. A typical measurement curve is shown in Fig. 4.9.

The residual resistance ratio (*RRR*) can be directly derived from this curve and serves as an index of the purity and overall quality of a sample. It is defined as the ratio between the resistance at 300 K and the resistance at 100 K: $RRR = R(300$ K$)/R(100$ K$)$. A large *RRR* is associated with a pure sample.

The zero-resistance critical temperature T_{c0} was defined as the temperature at which the resistance reaches 1% of the normal state resistance whereas the latter is determined right above the transition (see Fig. 4.10).

The third measure to determine the quality of a superconducting sample is the transition width. It is defined as the temperature difference between 0.9 and 0.1 R_n (see Fig. 4.10).

The *RRR* value for the 50 nm thick YBCO in dependence on the CeO_2 buffer layer thickness is displayed in Fig. 4.11(a). For decreasing CeO_2 thickness the *RRR* value increases reaching a value of 2.7 for CeO_2 thicknesses $\leq$ 30 nm. The critical temperature of the deposited samples is shown in Fig. 4.11(b). By the reduction of the CeO_2 thickness from 50 to 8 nm the critical temperature

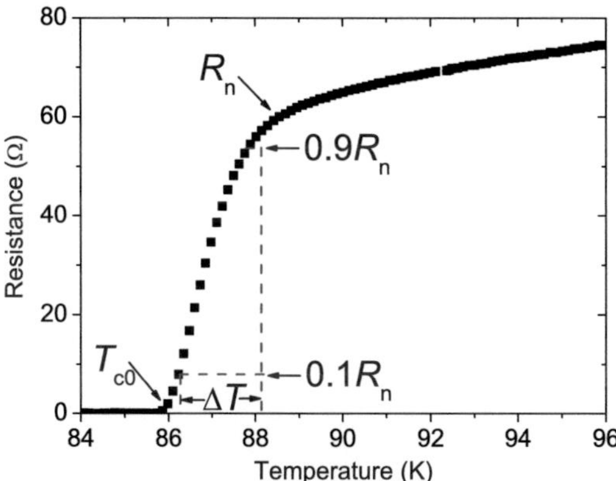

Fig. 4.10.: Zoom of the superconducting transition with the definitions of the normal state resistance R_n, the zero-resistance critical temperature T_{c0} and the transition width ΔT.

was improved by ≈ 3 K reaching 85.9 K for the 8 nm thick CeO_2 buffer layer. The dependence of the transition width ΔT of the 50 nm thick YBCO film on the CeO_2 thickness is shown in Fig. 4.12. The transition width decreases with decreasing buffer layer thickness reaching $\Delta T = 1.7$ K for the 8 nm thick CeO_2 buffer layer.

In summary, the 8 nm thick CeO_2 layer lead to a significant improvement of the superconducting YBCO thin film compared to the 50 nm thick buffer layer. Due to a shorter deposition time, the thinner CeO_2 films have less droplets leading to a smoother film surface. The YBCO films grow more homogeneous on these surfaces which results in an improvement of the crystalline quality.

For the development of radiation detectors a broad range of superconducting film thicknesses is required. Thus, YBCO film thicknesses between 10 and 100 nm were deposited on top of the optimized 8 nm thick buffer layer and compared to the 50 nm thick buffer layer earlier used. The dependence of T_{c0} on the YBCO film thickness for the 8 and 50 nm thick CeO_2 buffer layer is shown in Fig. 4.13. For YBCO film thicknesses above 30 nm T_{c0} is nearly constant and well

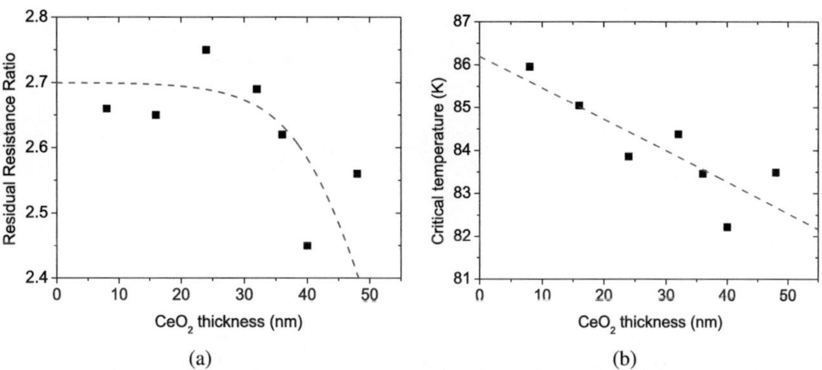

(a)

(b)

Fig. 4.11.: Dependence of the residual resistance ratio (a) and the critical temperature (b) of the 50 nm thick YBCO film on CeO_2 with thicknesses from 8 to 50 nm. The dashed lines are to guide the eye.

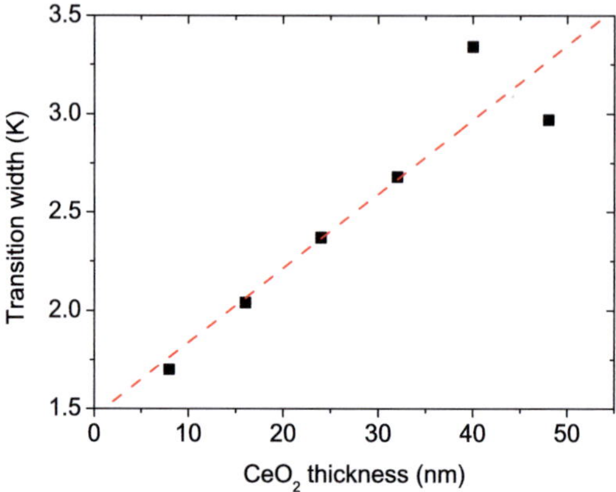

Fig. 4.12.: Dependence of the YBCO transition width ΔT on the CeO_2 thickness. With decreasing CeO_2 thickness the transition width decreases reaching $\Delta T = 1.7$ K for the 8 nm thick CeO_2 buffer layer. The dashed line is to guide the eye.

above the liquid nitrogen temperature of 77 K, reaching the optimized critical temperatures T_{c0} of 86 K. For films thinner than 30 nm T_{c0} decreases significantly. This can be explained by the aforementioned lattice mismatch leading to oxygen deficiency in the YBCO film, which results in a reduction of T_{c0} [121] down to 25 K for 15 nm thick films. In particular, the T_{c0} of films with $d < 30$ nm has been improved by the decrease of the CeO_2 layer thickness (see Fig. 4.13). At the CeO_2 film thickness of 8 nm an increase of T_{c0} up to 74 K for the 18 nm thick YBCO film was achieved.

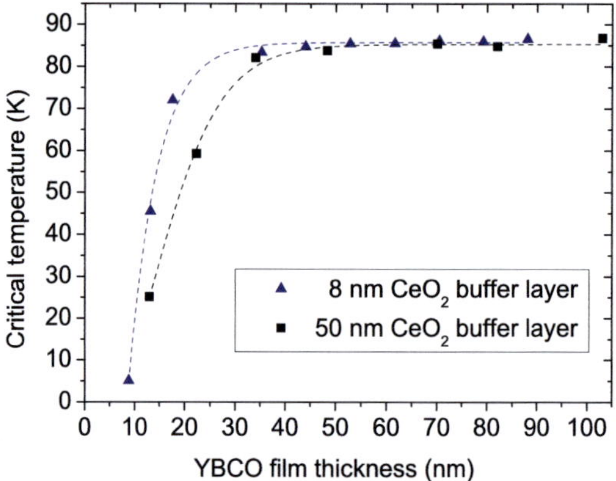

Fig. 4.13.: Dependence of the zero-resistance critical temperature on the YBCO film thickness for as-deposited films with different buffer layers: 50 nm thick CeO_2 buffer layer (squares), 8 nm thick CeO_2 buffer layer (triangles). The dashed lines are to guide the eye.

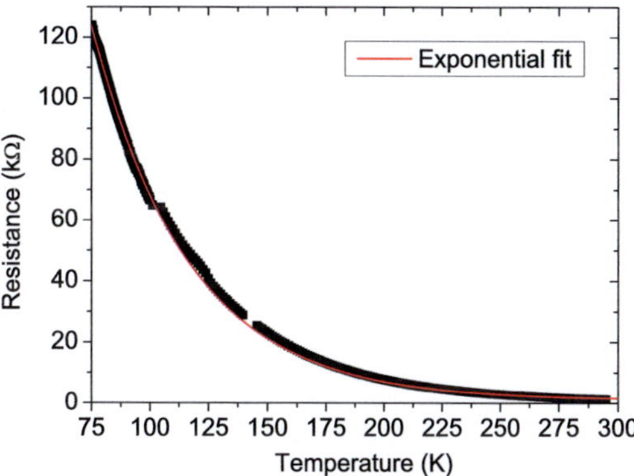

Fig. 4.14.: Temperature dependence of the resistance of a 2×10 mm^2 stripe of PBCO. The exponential increase of resistance with decreasing temperature of the semiconducting PBCO is emphasized by the exponential fit (solid line).

4.2.3. PBCO layer optimization

To protect the YBCO thin film against degradation by humidity in the atmosphere and the patterning process, a PBCO protection layer was deposited on top of the YBCO film. PBCO is a semiconducting material which shows an exponential increase in resistance with decreasing temperature (see Fig. 4.14). The thickness of the PBCO protection layer has been optimized to be thick enough to guarantee good protection against degradation, but thin enough to allow oxygen diffusion through the PBCO during the annealing process so that fully oxidized YBCO thin films can be deposited.

In Fig. 4.15(a) the temperature dependence of the resistance for a 50 nm thick YBCO layer with different PBCO protection layer thicknesses between 14 and 35 nm is shown. For increasing PBCO thickness the protection layer resistance decreases which leads to a shunting of the YBCO layer

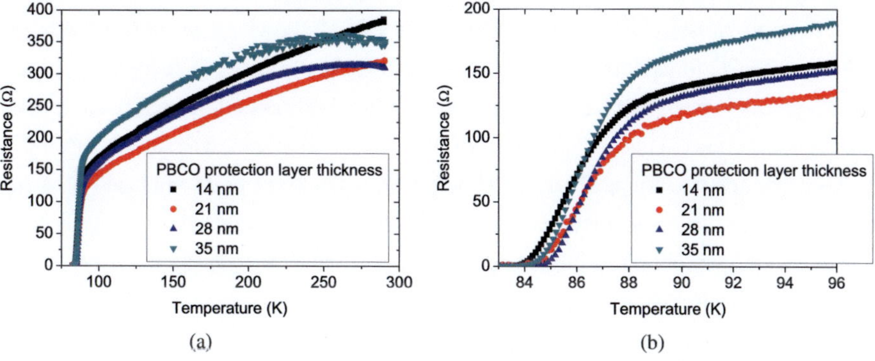

Fig. 4.15.: (a) Temperature dependence of resistance for the 50 nm thick YBCO for different thicknesses of the PBCO protection layer indicated in the legend. (b) Zoom to the superconducting transition revealing the highest transition temperatures for the 21 and 28 nm thick PBCO protection layer.

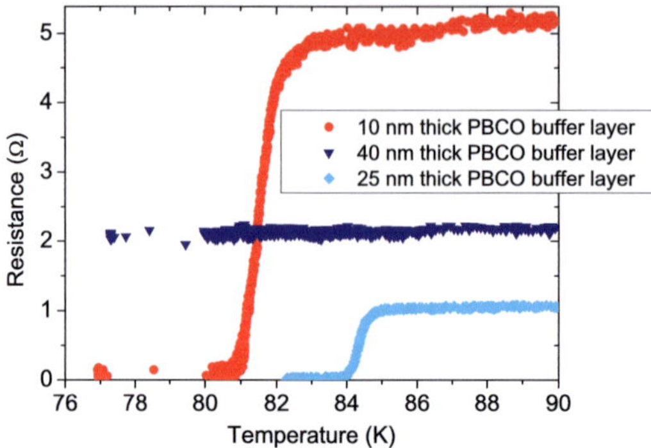

Fig. 4.16.: Temperature dependence of the resistance for a 20 nm thick YBCO layer for three different thicknesses of the PBCO buffer layer. The best results were achieved for a 25 nm thick PBCO buffer layer.

underneath resulting in a deviation from the linear temperature dependence at room temperature as shown in Fig. 4.15(a). However, for temperatures below 150 K the slopes of the curves are similar, since the resistance of the semiconducting PBCO becomes very large (see Fig. 4.14) compared to the YBCO layer. The zoom of the superconducting transition displayed in Fig. 4.15(b) shows a reduced critical temperature for the thickest PBCO protection layer of 35 nm. This can be explained by a reduced oxygen content in the superconducting YBCO underneath due to hampered oxygen diffusion through the thick PBCO layer. The highest zero-resistance transition temperature is achieved for the 28 nm thick layer with a slight reduction for the 21 nm thick layer. Since a minimum protection layer thickness is advantageous for the oxygen diffusion into the YBCO during the deposition process an optimum thickness for the PBCO passivation layer of 25 nm was determined and used in the following work.

The crystalline mismatch between the superconducting YBCO and the buffer layer of CeO_2 was further reduced by the introduction of an additional buffer layer made from PBCO. The lattice mismatch of approximately 1% between YBCO and PBCO allows for growing of very thin superconducting YBCO films with high crystalline quality. As discussed in section 4.2.2 the buffer layer thickness has to be optimized to be thick enough for good crystalline quality but thin enough to have a smooth surface. Thus, the thickness of the PBCO buffer layer was varied between 10 and 40 nm. The optimum was found for a PBCO buffer layer thickness of 25 nm (see Fig. 4.16). For thinner buffer layers the critical temperature was significantly reduced. For thicker buffer layers the same behaviour was observed showing even no superconducting transition for the 40 nm thick buffer layer as shown in Fig. 4.16.

The introduction of the 25 nm thick PBCO buffer layer combined with the PBCO protection layer of the same thickness resulted in a significant improvement of the YBCO superconducting properties. In Fig. 4.17 the zero-resistance critical temperature and the transition width in dependence on the YBCO film thickness is shown for the deposition process with the 8 nm thick CeO_2 buffer layer compared to the deposition process with the additional PBCO buffer and protection layers.

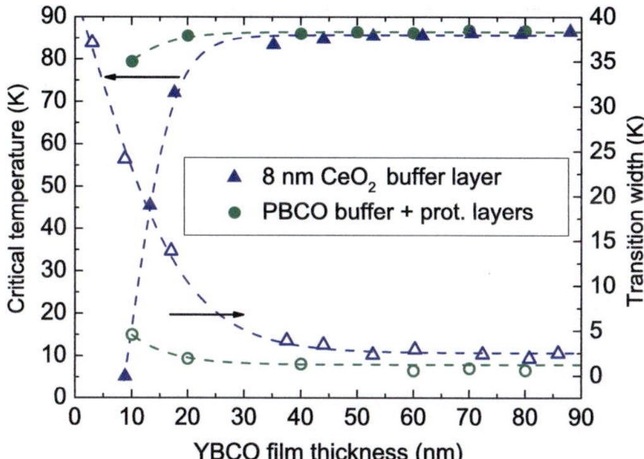

Fig. 4.17.: Dependence of zero-resistance critical temperature (closed symbols) and transition width (open symbols) on the YBCO film thickness for as-deposited films. The superconducting properties were significantly improved by the introduction of the 25 nm thick PBCO buffer and protection layers (circles) compared to the 8 nm thick CeO$_2$ buffer layer only (triangles). The dashed lines are to guide the eye.

In particular for films thinner than 30 nm the superconducting characteristics were significantly improved with an increase of ≈ 60 K in the critical temperature for the 10 nm thick YBCO film. The transition width was reduced by a factor of 5 from 25 K to 5 K for the 10 nm thick film. The optimized superconducting properties of our YBCO thin films are summarized in Fig. 4.18. For a wide YBCO thickness range between 10 and 100 nm high-quality YBCO thin films have been realized reaching critical temperatures of 86.5 K well above the liquid nitrogen temperature. Even for the thinnest YBCO film of only 10 nm thickness that corresponds to only eight unit cells a critical temperature T_{c0} of 79.4 K was achieved (see Fig. 4.18) allowing device operation of even the thinnest samples with liquid nitrogen.

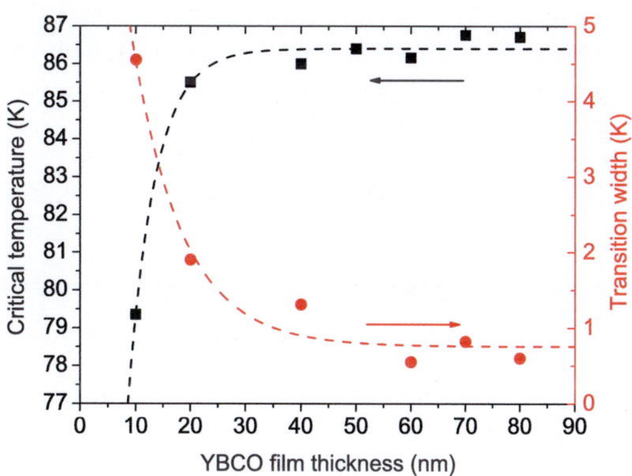

Fig. 4.18.: Dependence of zero-resistance critical temperature (squares) and transition width (circles) on the YBCO film thickness for as-deposited films with 8 nm thick CeO$_2$ and 25 nm thick PBCO buffer layers covered by a 25 nm thick PBCO layer for protection and a 140 nm thick Au layer to contact the film [PSR$^+$12]. The dashed lines are to guide the eye.

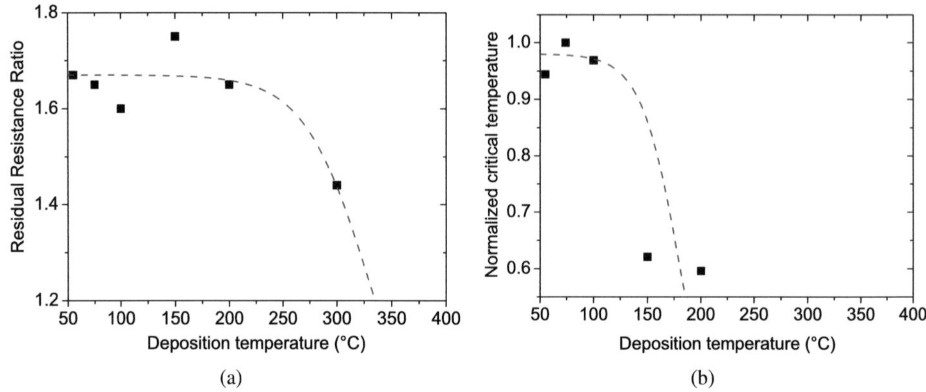

Fig. 4.19.: (a) Dependence of the residual resistance ratio (*RRR*) of a 30 nm thick Au layer on the deposition temperature [Wer10]. The *RRR* value decreases for deposition temperatures above 200°C. (b) Dependence of the critical temperature normalized to the maximum of the 15 nm thick YBCO film on the deposition temperature of the Au layer [Wer10]. The critical temperature decreases significantly with increasing deposition temperature showing no transition above 200°C. The dashed lines are to guide the eye.

4.2.4. Au layer optimization

For the development of sensitive THz detectors, an antenna is required to couple the incoming radiation to the small detecting element. A typical material for the realization of THz antennas is gold (Au) due to its low losses at high frequencies and corrosion-resistance. The *in-situ* Au deposition is an essential advantage for high-frequency THz devices, since losses due to a contact resistance between the Au antenna and the detecting element can be significantly reduced. For this reason, the *in-situ* Au deposition using pulsed-laser deposition has been developed and optimized [Wer10]. The main challenge for the PLD Au deposition are the high energy densities between 2 and 4 J/cm^2 required due to the high UV light reflectivity of Au. Thus, the Au characteristics in dependence of laser energy density were studied. The second parameter of the optimization was the deposition temperature.

A 30 nm thick Au layer was deposited directly onto the CeO$_2$-YBCO bilayer without any additional PBCO layer. The Au deposition temperature was varied between 50 and 400°C. In Fig. 4.19(a) the dependence of the residual resistance ratio *RRR* for the 30 nm thick Au layer on the deposition temperature is shown. Above 200°C the *RRR* value decreases significantly, showing even insulating behaviour at a deposition temperature of 400°C.

While the *RRR* value of the Au layer showed only a reduction above 200°C, the critical temperature of the superconducting YBCO film (15 nm thickness) underneath showed a much stronger dependence. Above 100°C the critical temperature decreased already to 60% of the critical temperatures of the YBCO films deposited between 50 and 100°C as shown in Fig. 4.19(b). Above 200°C the YBCO film shows no superconducting transition any more. This can be explained by diffusion of Au atoms in the YBCO crystalline lattice destroying superconductivity. Thus, the op-

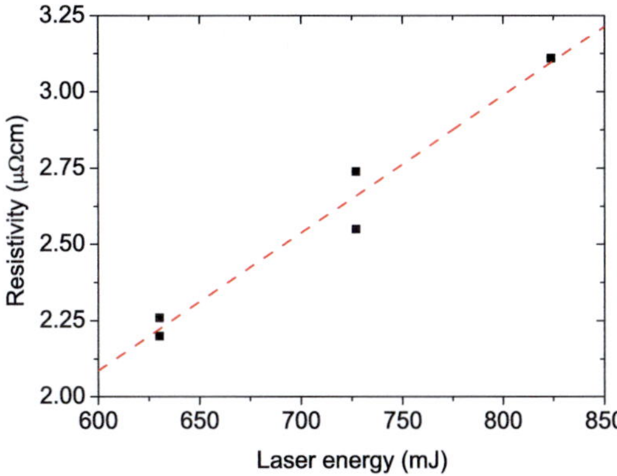

Fig. 4.20.: Room-temperature resistivity of Au in dependence of laser energy [Wer10]. The resistivity decreases with decreasing laser energy. The dashed line is to guide the eye.

timum deposition temperature for the Au layer was determined to 75°C, which does not suppress the critical temperature of the superconducting film underneath and results in a *RRR* value of a 30 nm thick Au layer of 1.7 (Au bulk value $\approx$ 3.5 [137]).

Beside the deposition temperature, also the influence of the laser energy was studied. In Fig. 4.20 the Au room-temperature resistivity in dependence of the laser energy is shown. The measured resistivity values are for the laser energy of 725 mJ in good agreement with literature values (ρ_{Au} = 2.44 $\mu\Omega$cm). With decreasing laser energy a decrease in the resistivity values is observed (see Fig. 4.20). Since a low resistivity is advantageous for high-frequency applications a laser energy of 630 mJ was chosen resulting in an energy density on the Au target of $\approx$ 3 J/cm^2 and a deposition rate of 0.15 nm/s.

In summary, the pulsed-laser deposition of YBCO thin films of 10 to 100 nm thickness embedded in a multilayer structure on sapphire has been optimized. High-quality superconducting properties have been achieved allowing for device fabrication and material studies. For a typical YBCO detector thickness of 30 nm a zero-resistance critical temperature of 86 K was achieved with a transition width below 1 K and a normal state resistivity of 850 $\mu\Omega$cm right above the superconducting transition.

4.3. YBCO thin films in magnetic fields

After the optimization and characterisation of the YBCO thin films embedded in a multilayer structure, the superconducting thin films were characterized in magnetic field to analyze the Ginzburg-Landau coherence length as well as the electron diffusion coefficient of our YBCO thin films. The Ginzburg-Landau coherence length is a direct measure for the vortex core diameter in our YBCO thin films [138] which allows us to determine the minimum lateral dimension of a YBCO detector element in which a vortex can exist. The electron diffusion coefficient is a measure of the defect

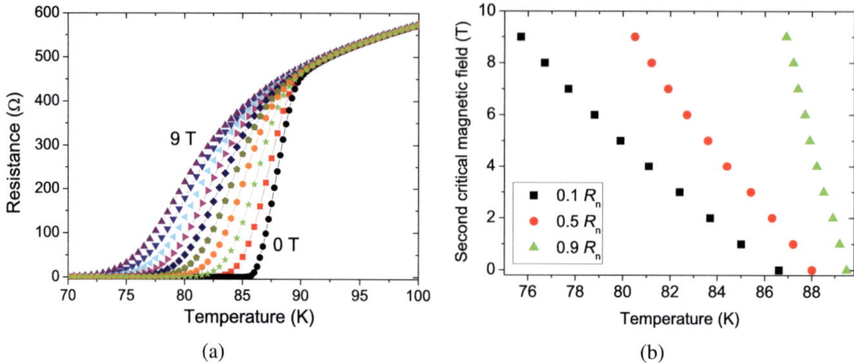

Fig. 4.21.: (a) Temperature dependence of the resistance of the 50 nm thick YBCO film in magnetic fields with inductions from 0 through 9 T plotted in steps of $\Delta B = 1$ T [PIE$^+$12]. The lines are guides for the eye. (b) Temperature dependence of the second critical magnetic field for different criteria of the critical temperature: $T(0.1R_n)$, $T(0.5R_n)$ and $T(0.9R_n)$.

concentration due to its direct proportionality to the electron mean free path [139]. The defects act as potential vortex pinning centers inside the YBCO thin films. Thus, by studying the electron diffusion coefficient in dependence of the YBCO film thickness we can derive an indirect measure for relative changes in the pinning center concentration within our detectors.

For this, YBCO thin films with thicknesses between 5 and 100 nm were deposited on sapphire substrate with CeO_2 and PBCO buffer layers and covered by the PBCO protection layer. Magnetoresistivity of these samples has been measured in the Physical Property Measurement System at the University of Zurich to analyze the temperature dependence of the second critical magnetic field B_{c2} as discussed in [PIE$^+$12], which is required to determine the coherence length and the electron diffusion coefficient. Since the criterion used for the definition of the transition temperature strongly influences on the slope of $B_{c2}(T)$ and thus on the electron diffusion coefficient, the transition temperature definition according to the Aslamazov-Larkin-theory is introduced in section 4.3.1. Due to the lack of a high-temperature superconducting theory for YBCO, the Ginzburg-Landau (GL) theory for low-temperature superconductors will be applied in the following chapter. Observed experimental deviations from this theory are discussed in section 4.3.2. Finally, in section 4.3.3 the Ginzburg-Landau coherence length and the electron diffusion coefficient are discussed.

4.3.1. The second critical magnetic field - determination of the transition temperature

Fig. 4.21(a) shows magnetoresistivity data near the superconducting transition for the 50 nm thick sample measured in magnetic fields up to 9 T parallel to the c-axis of the YBCO film. For all film thicknesses between 5 and 100 nm a broadening of the transition with the increase of magnetic field was observed. From this data, the second critical magnetic field can be directly determined.

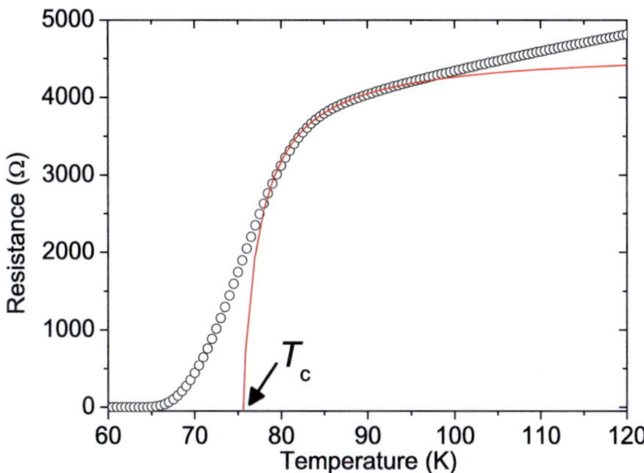

Fig. 4.22.: Temperature dependence of the resistance for the 8 nm thick sample (circles). The solid line shows the best fit of the superconducting transition with the AL theory for two-dimensional superconducting systems [PIE+12].

However, the absolute value derived from this data depends strongly on the criterion used for the definition of the transition temperature.

Fig. 4.21(b) shows the temperature dependence of the second critical magnetic field for three different criteria of the transition temperature: $T(0.1R_n)$, $T(0.5R_n)$ and $T(0.9R_n)$. For the same temperature the second critical magnetic field differs strongly depending on the transition temperature criterion. Therefore, it was important to define the correct transition temperature for all films. Thus, the mean-field transition temperature T_c was evaluated by fitting the resistive transition in zero magnetic field with the fluctuation conductivity theory of Aslamazov and Larkin (AL) [140]. Expressed in terms of the measured resistance R and system dimensionality $m = 1, 2, 3$ it renders

$$R(T) = \frac{R_n}{1 + CR_n\frac{A}{l}A_m\left(\frac{T_c}{T-T_c}\right)^{2-0.5m}} \tag{4.1}$$

with A_m a constant factor depending on the dimensionality of the studied superconducting system [138]. R_n is the normal state resistance at the onset of the transition in zero field, A is the cross-section of the samples and C is a fitting parameter. The $R(T)$-dependencies were fitted using equation 4.1 in a narrow temperature range in the vicinity of the onset of the superconducting transition using all three values for m. The model for two-dimensional (2D) superconducting systems ($m = 2$) with $A_2 = e^2/(16\hbar d)$ fit our data best with the smallest values of C. Here $\hbar$ is the reduced Planck constant and e is the elementary charge. The obtained results are in agreement with earlier reported data [141] and our expectations, since we consider temperatures very close to T_c where ξ becomes large thus justifying the two-dimensional approximation $d \ll \xi$ [138]. A typical result is shown in Fig. 4.22 where the temperature dependence of the resistance (circles) of the 8 nm thick YBCO film is fitted by the AL theory according to equation 4.1 (solid line) with $T_c = 75.3$ K indicated by the arrow. For all film thicknesses the extracted T_c values are close to the

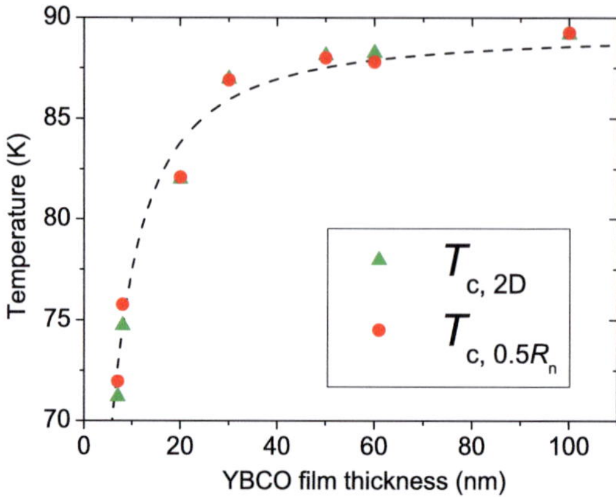

Fig. 4.23.: Dependence of the transition temperature on the YBCO film thickness according to the AL mean-field theory (triangles) and the $T(0.5\,R_n)$ criterion. The data shows very good agreement. The dashed line is to guide the eye.

midpoint of transition as demonstrated in Fig. 4.23. Therefore, we have chosen the 0.5 R_n-level of the transition to define the temperature T at which the externally applied magnetic field equals $B_{c2}(T)$.

4.3.2. Deviation of $B_{c2}(T)$ from Ginzburg-Landau-theory

At temperatures in the vicinity of T_c the temperature dependence of the second critical magnetic field $B_{c2}(T)$ is described as

$$B_{c2}(T) \propto (1 - T/T_c)^n \tag{4.2}$$

with $n = 1$ for conventional low-T_c superconductors according to the Ginzburg-Landau (GL) theory [142]. As discussed in the section above, the transition temperature was defined as $T_c = T(0.5\,R_n)$ to determine the second critical magnetic field.

Fig. 4.24 shows the series of the B_{c2}-temperature dependence for film thicknesses from 7 to 100 nm. The solid lines represent fits according to equation 4.2 using n as a fitting parameter. At temperatures $T < 0.95\,T_c$ (magnetic fields above 3 T) all films show a linear dependence of B_{c2} on temperature in agreement with the GL theory for conventional superconductors [142] (equation 4.2 with $n = 1$). In Fig. 4.24, this is emphasized by linear fits (dashed lines) to the data at large magnetic fields. At $T > 0.95\,T_c$ the $B_{c2}(T)$ dependences vary strongly with film thickness. We observe a downturn curvature of $B_{c2}(T)$ close to T_c for ultra-thin films with thicknesses between 7 and 20 nm with a minimum exponent of $n = 0.8$ for the 7 nm thin sample (solid line in Fig. 4.24). The exponent $n = 0.5$ has been observed by Fang et al. [143] for twinned, 150 µm sized YBCO crystals. The exponent n increases with the increase in film thickness converging to the linear GL dependence ($n = 1$) for $d = 30$ nm. The films thicker than 30 nm reveal an upturn curvature of $B_{c2}(T)$. Similar $B_{c2}(T)$ behavior has been reported by Tinkham [144] with an exponent of $n = 1.5$ for 100 µm YBCO crystals. Also Oh et al. [141] reported an upturn curvature

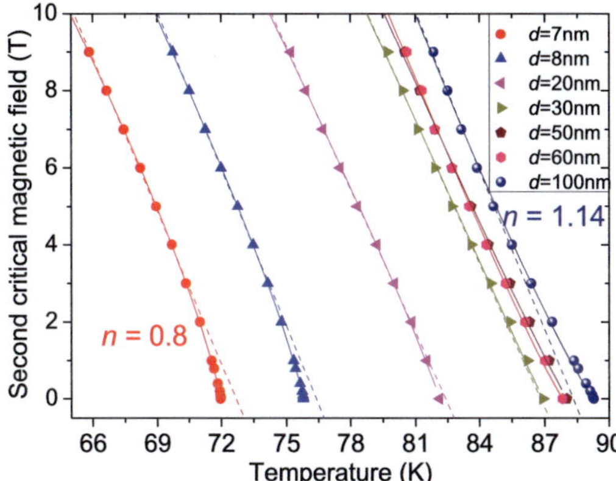

Fig. 4.24.: Temperature dependence of second critical magnetic field for YBCO thin films with different thicknesses indicated in the legend [PIE$^+$12]. The dashed lines are the low-temperature fits by equation 4.2 with $n = 1$. The solid lines show the best fit of the measurement data according to equation 4.2 with variable n.

with $n = 1.3$ for 1 µm thick epitaxial YBCO films using both a 0.9 R_n criterion and the critical fluctuation theory. The thickest studied YBCO film in this work with $d = 100$ nm shows a $B_{c2}(T)$ dependence with an exponent $n = 1.14$. Since the activation energy for flux motion is inversely proportional to the applied magnetic field a possible explanation for the observed behavior may be a change in the strength of vortex pinning in films with different thicknesses. However, since the excitation current during all magnetoresistivity measurements was 100 nA (corresponding to $j = 70$ A/cm^2 for $d = 7$ nm) influence of vortex depinning may be neglected.

4.3.3. $\xi(0)$ and D

With the clear definition of the transition temperature and the temperature derivative of the second critical magnetic field, the zero-temperature second critical magnetic field $B_{c2}(0)$, the coherence length at zero temperature $\xi(0)$ and the diffusion coefficient D can be derived from the magnetoresistivity data. For the following discussion of $B_{c2}(0)$, $\xi(0)$ and D the linear temperature dependence of the second critical magnetic field (see section 4.3.2 for $T < 0.95\,T_c$) was taken into account. A linear extrapolation of the measured $B_{c2}(T)$ down to $T = 0$ K certainly overestimates the real upper critical field at zero temperature. A more realistic value in the dirty limit is according to [145]

$$B_{c2}(0) = 0.69 T_c \left[\frac{dB_{c2}(T)}{dT} \right]_{T=T_c}. \tag{4.3}$$

In Fig. 4.25 the zero-temperature second critical magnetic field in dependence of the YBCO film thickness is shown. For film thicknesses above 30 nm $B_{c2}(0)$ is nearly constant with values of ≈ 80 T which is in good agreement with earlier published results [146]. For films thinner than 30 nm $B_{c2}(0)$ decreases down to 47 T for the 5 nm thick film. Since $B_{c2}(0)$ is directly proportional to the critical temperature, this behaviour is in agreement with the zero-resistance critical tempera-

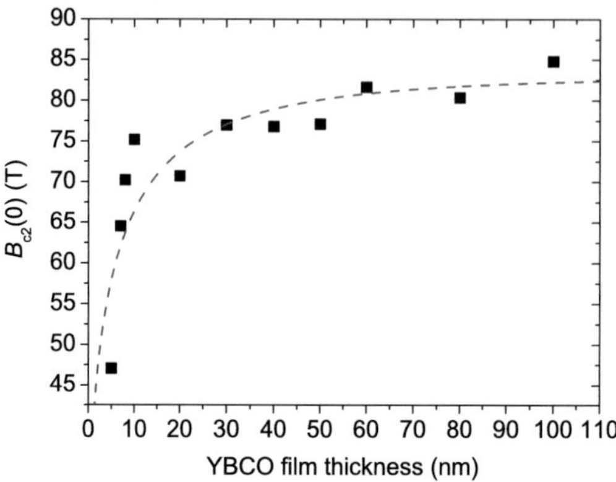

Fig. 4.25.: Zero-temperature second critical magnetic field in dependence of the YBCO film thickness. The dashed line is to guide the eye.

ture data where a similar suppression of the superconducting characteristics has been observed for the film thicknesses below 30 nm (see section 4.2). The resulting upper critical fields of 50 T and larger suggest that $B_{c2}(0)$ is further reduced due to spin paramagnetism and spin-orbit scattering [147]. The obtained value should thus be taken as the upper limit for $B_{c2}(0)$.

With the $B_{c2}(0)$ values the GL coherence length at $T = 0$ K can be determined according to the GL theory, where the upper critical magnetic field is connected with the superconducting coherence length ξ and the magnetic-flux quantum $\Phi_0 = h(2e)^{-1}$:

$$B_{c2}(T) = \frac{\Phi_0}{2\pi\xi(T)^2}. \tag{4.4}$$

However, due to the uncertainty in $B_{c2}(0)$, this value is in turn a lower limit. However, because of the square-root dependence of ξ on B_{c2}, this should give a satisfactory estimate of the coherence length in our films. Fig. 4.26 shows the zero-temperature coherence length in dependence of the thickness of our YBCO thin films. For film thicknesses above 30 nm a coherence length of 2 nm was found which is in good agreement with earlier published results [146]. Below 30 nm the coherence length increases reaching 2.6 nm for the 5 nm thick film. These values of ξ suggest that the vortex core diameter in our YBCO thin films is much smaller than any dimension of our YBCO detector. This allows vortices to penetrate our YBCO detector element supporting the idea of a vortex-assisted detection mechanism (see section 6.4).

The equation of the electron diffusivity D for dirty superconductors according to [148] is

$$D[cm^2/s] = -\frac{4k_B}{\pi e}\left[\frac{dB_{c2}(T)}{dT}\right]^{-1}_{T=T_c} = -\frac{1.097}{\left[\frac{dB_{c2}(T)}{dT}\right]_{T=T_c}[T/K]} \tag{4.5}$$

where k_B is the Boltzmann constant and e the elementary charge. In Fig. 4.27 the dependence of the diffusion coefficient on the YBCO film thickness is shown. For the films between 7 and 80 nm

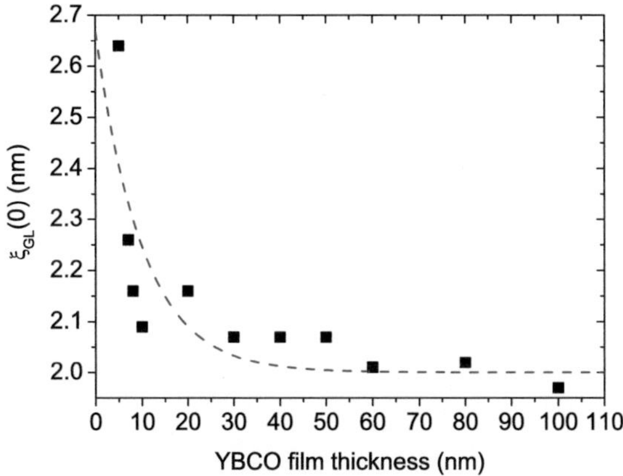

Fig. 4.26.: Zero-temperature coherence length in dependence of the YBCO film thickness. The dashed line is to guide the eye.

the diffusion coefficient shows only slight deviations from the mean value of 0.84 cm^2/s. Only the thinnest film thickness of 5 nm deviates to a higher value of 0.95 cm^2/s, and the thickest, 100 nm film shows a lower value of 0.8 cm^2/s.

Thus, we can conclude that the electron diffusion coefficient for our YBCO thin films is only slightly dependent on film thickness and is in the range of 0.84 cm^2/s (see Fig. 4.27). This shows that the electron mean free path and thus the defect concentration in our YBCO thin films does not change significantly with film thickness. Therefore, we can assume that also the pinning center concentration in our YBCO thin film detectors does not change significantly with film thickness.

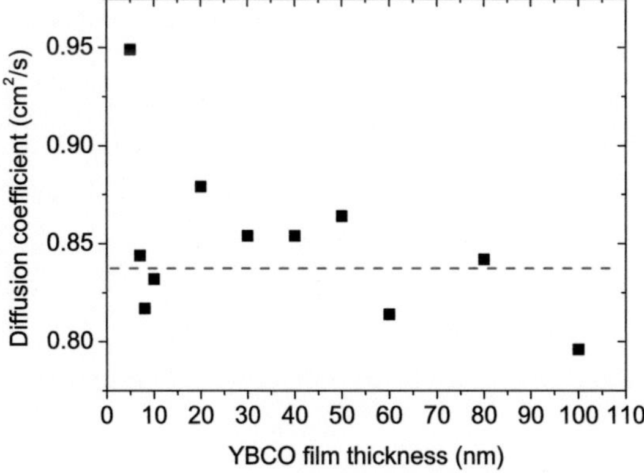

Fig. 4.27.: Dependence of diffusion coefficient on YBCO film thickness. The diffusion coefficient is nearly constant for all film thicknesses with a value of 0.84 cm^2/s. The dashed line is to guide the eye.

4.4. Conclusions

The pulsed-laser deposition setup used in this work for the fabrication of ultra-thin YBCO films of 10 to 100 nm thickness was introduced in section 4.1. In section 4.2 the optimization of the deposition process for the YBCO thin films on sapphire has been discussed in detail. By the introduction and optimization of a CeO_2 and PBCO buffer layer as well as a PBCO protection layer, high-quality superconducting YBCO thin films have been achieved. For a typical YBCO detector thickness of 30 nm a zero-resistance critical temperature of 86 K was achieved with a transition width below 1 K. These values are very close to earlier reported values for much thicker films of 100 nm on sapphire [121].

In section 4.3 the optimized YBCO thin films were studied in magnetic fields up to 9 T to get insight in the vortex physics occurring inside our superconducting thin films. The transition temperature in the YBCO thin films was determined by the model for two-dimensional superconducting systems according to Aslamazov and Larkin. For the extraction of the slope of $B_{c2}(T)$ the GL-theory was used which showed for temperatures very close to T_c ($T > 0.95\,T_c$) and low magnetic fields a deviation from the predicted linear behaviour which might be attributed to vortex depinning in thin YBCO films. The zero-temperature GL coherence length showed an increase with the reduction of the YBCO film thickness. However, with absolute values below 3 nm these dimensions are much smaller than the detecting element. This suggests that our YBCO thin film detectors are penetrated by magnetic vortices which is important for the discussion of the detection mechanism in our YBCO detectors in section 6.4. For the diffusion coefficient no dependence on the YBCO film thickness was found, indicating that the defect concentration and thus the pinning center concentration does not change significantly with a reduction of the film thickness.

5. YBCO thin-film detectors - Fabrication and DC characterization

The as-deposited YBCO thin films with high-quality superconducting characteristics as discussed in chapter 4 are now structured for radiation detector applications. In this chapter the patterning technology for YBCO detectors with film thicknesses between 15 and 50 nm with micrometer and sub-micrometer lateral dimensions is presented. The main challenge was the development of a sub-micrometer patterning process for the YBCO detecting element which is embedded in a several hundred micrometer large antenna structure and a millimeter-sized coplanar transmission readout. The DC characterisation of the fabricated devices is discussed regarding the critical temperature and critical current density and a study of the long-term stability is presented.

5.1. Patterning technology

To develop a device applicable from optical to THz wavelengths a thin-film YBCO bridge was embedded in a planar THz antenna. Whereas the optical radiation is directly absorbed in the YBCO bridge, the antenna is required to couple the THz radiation to the detecting element. In this section, the different antenna concepts used in this work are briefly presented. Details concerning the antenna simulations and design can be found in [149]. After the introduction of the THz antennas, the development of the technology for patterning of micrometer- and nanometer-sized YBCO bridges is discussed.

5.1.1. THz detector layout

Log-spiral antenna

To provide broadband coupling in the THz frequency range planar THz antennas were implemented. The standard layout was based on a logarithmic-spiral antenna which was designed by Dipl.-Ing. Alexander Scheuring [PSR$^+$12], [149]. The antenna was embedded into a coplanar readout line. A schematic layout of the antenna structure is shown in Fig. 5.1.

The antenna was designed with CST Microwave Studio [150] to couple radiation in the frequency range from 0.1 to 2 THz, which encompasses the spectrum of a typical CSR source [151, 152]. The impedance of the antenna was simulated for semi-infinite sapphire substrate. The real part is almost constant ($R_a \approx 65\ \Omega$) over the entire frequency range. The imaginary part is nearly zero [PSR$^+$12]. The simulated reflection parameter for a detector with a frequency-independent real impedance of 50 Ω is well below -10 dB from 0.15 THz up to more than 2.5 THz, which corresponds to a coupling efficiency higher than 90% between antenna and detector element in this frequency range [PSR$^+$12]. SEM images of a ready-fabricated device are shown in Fig. 5.2.

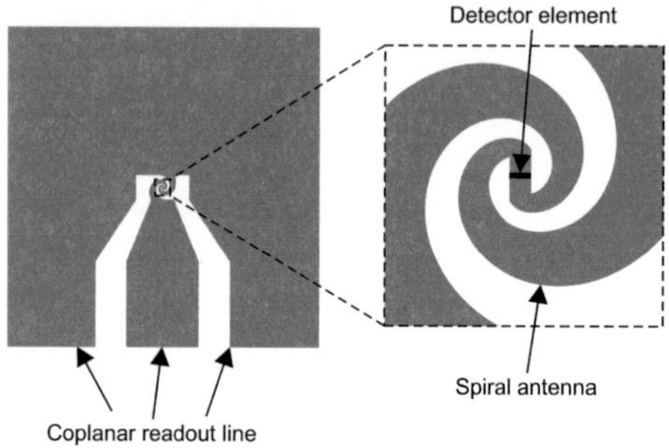

Fig. 5.1.: Left panel shows the sample layout. Gray color shows Au metallization. The log-spiral antenna is embedded into a coplanar readout line. Right panel shows the zoom-in of the antenna with the YBCO bridge (black) at the center feed point [PSR$^+$12].

Low-frequency log-spiral antenna

For typical bunch durations σ_z between 2 and 10 ps at ANKA, most of the CSR power is emitted in the low THz frequencies between 0.03 and 1 THz (see [1] and Fig. 7.12). Since the log-spiral antenna discussed above has a lower cut-off frequency of 150 GHz, a new antenna design with a lower cut-off was developed to maximize the CSR power absorption in the low THz frequencies. As a rule-of-thumb the antenna upper and lower cut-off wavelength can be estimated using the inner and outer diameter of the designed spiral. According to [153] the outer diameter D is the diameter of the smallest circle that encompasses the spiral structure. The inner diameter d is the diameter of the smallest circle at which the arms still obey the spiral equation. For the log-spiral antenna discussed above the inner and outer diameter result in 9.6 µm and 323 µm, respectively. According to [153] this results in a lower cut-off wavelength of $\lambda_{min} \approx 20d = 192$ µm (1.56 THz) and an upper cut-off wavelength of $\lambda_{max} \approx 6D = 1938$ µm (155 GHz) which shows in particular for the lower cut-off frequency reasonable agreement with the simulations in CST Microwave Studio [TSH$^+$12].

Thus, the lower cut-off of the antenna bandwidth was reduced by increasing the outer diameter of the new antenna design. The sample layout with antenna and coplanar readout is shown in the left

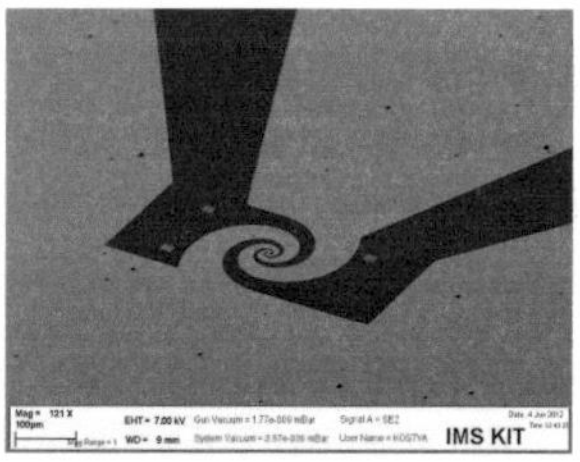

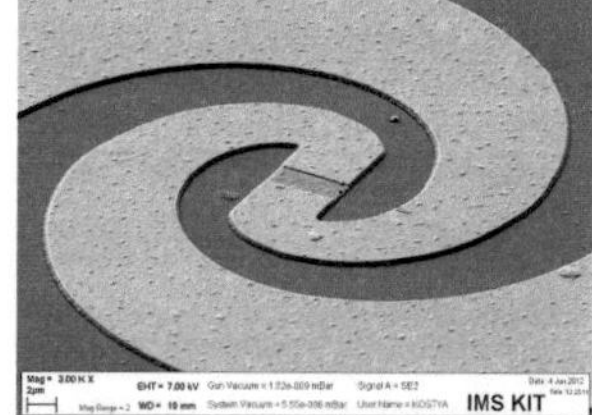

(a) Spiral antenna with coplanar readout

(b) Zoom of the antenna

(c) Zoom of the detecting element

Fig. 5.2.: SEM images of the log-spiral antenna design embedded in the coplanar readout line. The detecting YBCO bridge is located in the center of the antenna.

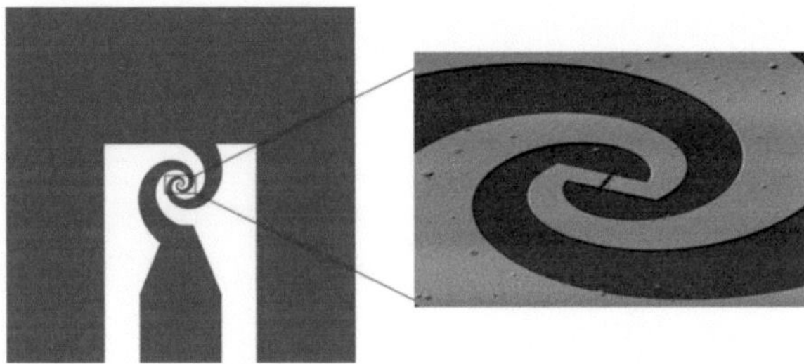

Fig. 5.3.: Left panel shows the sample layout. Gray color shows Au metallization. The log-spiral antenna with lower cut-off frequency is embedded into a coplanar readout line. Right panel shows a SEM image of the zoom of the antenna with the microbridge (black) at the center [TSH$^+$12].

panel of Fig. 5.3. The right panel shows a SEM image of the zoom of the detecting element. The inner diameter was designed to $d = 16$ μm, resulting in an upper cut-off frequency of 1 THz. The outer diameter was increased to 912 μm leading to a lower cut-off frequency of 55 GHz.

These results were crosschecked by the simulation of the antenna design [TSH$^+$12], [149] in CST Microwave Studio [150]. The reflection parameter was simulated for a detector real impedance of 50 Ω and is well below -10 dB from 0.07 THz to 1 THz which results in a coupling efficiency between antenna and detector of more than 90% in this frequency range.

5.1.2. Microbridges - Lithography and etching

An overview of the patterning process is schematically shown in Fig. 5.4 and discussed in [TRS$^+$13]. The as-deposited multi-layers (Fig. 5.4(a)) described in section 4.1.2 have been patterned by several electron-beam lithography and etching steps. For the electron-beam lithography the e-LiNE system from Raith GmbH was used whereas for the physical etching steps the rf ion source from the Nordiko ion milling system was employed.

In the first step, alignment marks were written with PMMA 950K 3% resist from Allresist GmbH. The PMMA resist was spun on the multilayer with 8500 rpm for 60 s (program C) resulting in a resist thickness of 180 nm. The sample was baked at only 150°C (standard is 175°C) for 5 minutes to reduce oxygen loss in the superconducting YBCO layer due to overheating. The settings for the electron beam lithography system were chosen to a high voltage of 5 kV and an aperture of 20 μm resulting in a beam current of $\approx$ 65 pA. The dose and area step size were chosen to 50 μC/cm^2 and 28 nm, respectively. The writing field was set to 100 μm. The resist was developed using the standard procedure of 30 s in the PMMA developer and 2 times 30 s in isopropanol.

The alignment marks were etched by argon ion milling. The acceleration voltage for the ions was chosen to 200 V at the positive grid and 270 V at the negative grid. The rf-power was adjusted to about 130 W to keep the current at the positive grid constant during the etching procedure at 64 mA. During etching the sample was cooled with a water chiller set to 10°C to reduce oxygen loss in the YBCO layer during the ion milling. The overall etching time was 20 min while the shutter was closed every 30 s to avoid overheating (10 min effective etching time). By this the

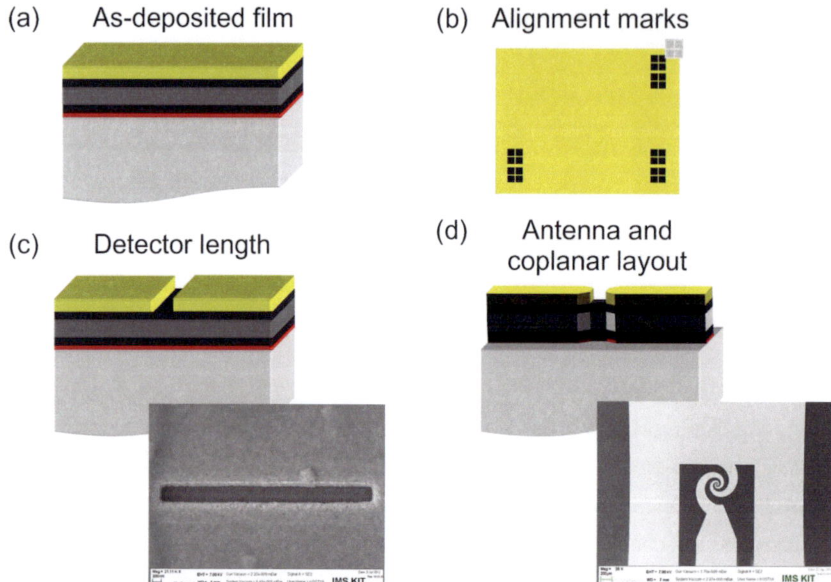

Fig. 5.4.: Patterning process for YBCO detectors: (a) The as-deposited multilayer CeO$_2$-PBCO-YBCO-PBCO-Au. (b) Top view of the patterned local alignment marks used for the alignment of the detector slit and the antenna layout. (c) Schematic and SEM image of the etched detector slit defining the final detector length. (d) Schematic of the central part of the detector structure showing the definition of the antenna arms and the width of the YBCO detector. SEM image of the final detector structure with the planar antenna embedded in the coplanar readout.

140 nm thick Au layer was etched through, resulting in an etching rate of 14 nm/min. The top view of the YBCO sample after the etching of the alignment marks is schematically shown in Fig. 5.4(b). A set of three local alignment marks is used for the alignment of the detector slit and the antenna, respectively, as described below.

To prepare the sample for the next patterning step of defining the lengths of the detecting microbridges, one has to make sure that the PMMA resist is completely removed. For this, the sample was cleaned with acetone and isopropanol followed by a 2 min treatment of O$_2$ plasma ashing.

After this the same 180 nm thick PMMA resist was spun on the sample and baked at 150°C for 5 min. The high voltage was set to 5 kV and an aperture of 10 µm resulting in a beam current of $\approx$ 20 pA. The area step size was 30 nm with a dose of 45 µC/cm^2. The writing field was set to 500 µm. The development procedure was as described above. To make sure that no resist was left in the microbridges and to reduce the etching time in the isotropic wet-etching, the sample was etched with an ion milling process for an effective etching time of 10 min (20 min: 30 s opened, 30 s closed shutter). The positive acceleration voltage was 120 V, while the negative voltage was kept at 270 V. The rf-power was adjusted to about 130 W to get a constant current at the positive grid of 42 mA. After this the sample was etched isotropically in an I$_2$KI-solution diluted with distilled water (1 : 10) for about 60 s until the microbridges were opened (see Fig. 5.4(c)). The

Fig. 5.5.: AFM image of a ready-fabricated micrometer sized detecting element embedded in the THz antenna made from a Au thin film.

etching process was stopped by isopropanol. A typical under-etching of the resist of 0.5 µm was taken into account.

The sample was cleaned again using acetone, isopropanol and 2 min of O_2 plasma ashing.

To pattern the antenna and coplanar readout negative e-beam resist (ARN 7520.18 from Allresist GmbH) was spun on the sample resulting in a resist thickness of 350 nm. The sample was baked at 85°C for 2 min. The high voltage was set to 7 kV with a writing field of 500 µm. For the antenna writing the aperture was chosen to 20 µm with an area step size of 20 nm and a dose of 60 µC/cm². The beam current was $\approx$ 75 pA. For the coplanar readout the working distance was enlarged from 6 to 10 mm and the high current mode was used with an aperture size of 120 µm. This resulted in a beam current of $\approx$ 5.5 nA. The area step size was 200 nm with a dose of 195 µC/cm². The resist was developed in a mixture of MIF and distilled water (3 : 1) for 1 min. The development was stopped by distilled water.

The final etching step was realized by argon ion milling. Since PBCO and YBCO are both very hard materials (ceramic-like), the parameters for the ion milling etching were optimized towards high acceleration voltages. The acceleration voltage was increased to 250 V with the negative voltage remaining at 270 V. The rf-power was adjusted to about 170 W to have a positive current of 64 mA. For a 15 nm thick YBCO layer, with 25 nm thick PBCO buffer and protection layers and the 140 nm thick Au layer, an etching time of 60 min was required, resulting in an average etching rate of 3.4 nm/min. The center part of the final device as well as a SEM image of a YBCO detector embedded in Au antenna and coplanar readout is shown in Fig. 5.4(d).

In Fig. 5.5 an atomic force microscopy (AFM) image of a microbridge embedded in the Au THz antenna is shown. The droplets typical for PLD are clearly visible on the Au surface. Typical

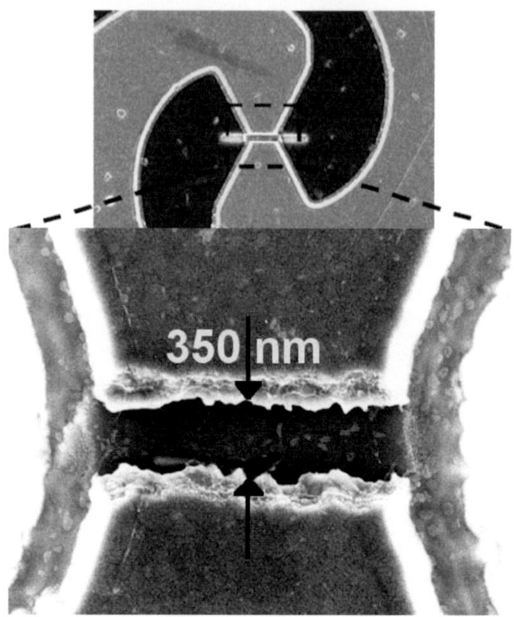

Fig. 5.6.: SEM images of the center part of the antenna, with a YBCO detecting element of 350 nm length and 1.5 µm width [TRS+13].

dimensions of the YBCO microbridges for a 30 nm film thickness were a width of 4.7 µm and a length of 2.5 µm.

5.1.3. Nanobridges - Lithography and etching

The main limitation for the reduction of the length of the detecting element to sub-micrometer dimensions, was the strong under-etching of ≈ 0.5 µm in the I_2KI-solution of the patterning process described above. To overcome this limitation the pre-etching in the Ar ion-milling system of the Au cap layer was extended to reduce the etching time in the I_2KI-solution and thus the under-etching.

For the extension of the ion-milling etching procedure a thicker resist was required. Thus, for patterning of the detecting elements PMMA 950K 5% resist from Allresist GmbH with a thickness of 460 nm was used. The resist was baked for 5 min at 150°C.

For patterning with electron-beam lithography the acceleration voltage was increased to 15 kV. This allowed to increase the steepness of the edges in the resist and thus, a correct reproduction of the design in the resist. The aperture and writing field were kept at 10 µm and 500 µm, respectively. The area step size and the dose were set to 28 nm and 90 µC/cm^2, respectively. The resist was developed using the standard procedure for PMMA resist described above.

The samples were etched in the Ar ion milling system to remove most of the Au layer. The remaining Au layer in the detecting slits after ion-milling etching of ≈ 30 nm was then removed by wet etching using the I_2KI-solution. To further reduce the under-etching, the I_2KI-solution was

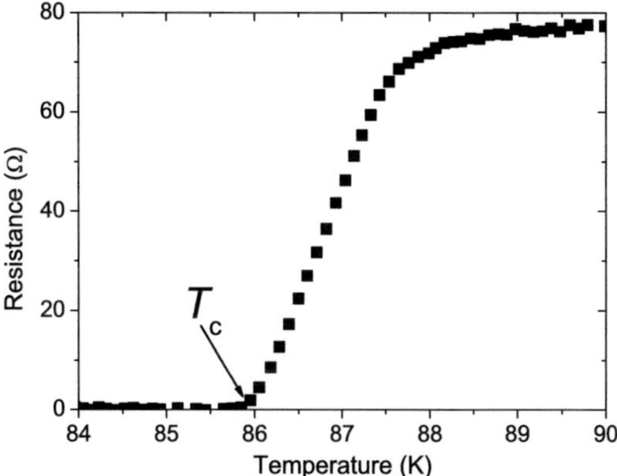

Fig. 5.7.: Temperature dependence of the resistance for a 30 nm thick YBCO detector with a width of 4.7 µm and a length of 2.5 µm. The transition temperature is defined as the temperature where the resistance equals zero (see arrow).

strongly diluted using a mixture of I_2KI to distilled water of 1 : 12. The etching time remained constant compared to section 5.1.2 at about 60 s which is explained by the reduced etching rate of the diluted etchant (1 : 12). By this, the under-etching could be reduced to less than 80 nm and detecting elements as short as 300 nm were successfully realized. Fig. 5.6 shows the center part of a final device with a YBCO nanometer-sized detecting bridge embedded in the Au antenna.

5.2. DC characterization

To evaluate the influence of the patterning process on the as-deposited YBCO thin films as well as the quality of the patterned devices, the detectors were characterized in terms of their superconducting characteristics. Beside the determination of the critical temperature and critical current density after the detector fabrication also a long-term study regarding the stability of these parameters has been performed.

5.2.1. Critical temperature and critical current density

The superconducting properties of the fabricated detectors were characterized by a quasi four-probe measurement configuration. This means that on the detector chip the voltage and current lines were bonded to the same pads, however the measurement cables were implemented in a true four-probe measurement configuration. The temperature dependence of the resistance was measured to determine the zero-resistance critical temperature of the devices. In Fig. 5.7 a typical measurement curve is displayed for a 30 nm thick microbridge detector. The zero-resistance critical temperature was determined to 86 K. In Fig. 5.8 the dependence of the critical temperature on the YBCO film thickness is shown (squares) keeping the lateral dimensions of the microbridges constant ($w = 4.7$ µm, $l = 2.5$ µm). With the reduction of the YBCO film thickness the critical temperature decreases as discussed already in section 4.2. The comparison with the as-deposited films

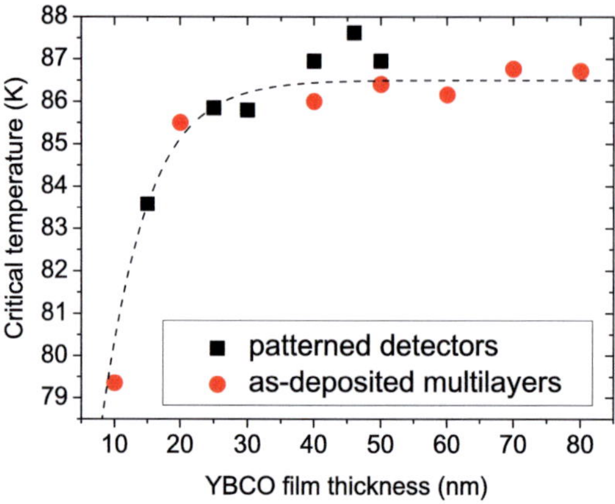

Fig. 5.8.: Dependence of the critical temperature on the YBCO film thickness for the as-deposited mul-
tilayers and patterned detectors. No degradation of the patterned samples in comparison to the
as-deposited films is observed.

(circles) shows no reduction of the critical temperature of the patterned samples. This allows the
conclusion that by the developed patterning process the surface and edges of the superconducting
material are not affected to an extend measurable in micrometer-sized devices.

Another parameter to determine the superconducting characteristics of the patterned detectors is
the critical current density. For this, the current-voltage (IV) characteristics for all detectors have
been determined at 77.36 K in liquid nitrogen. A typical measurement curve is shown in Fig. 5.9
for a 30 nm thick microbridge. The s-shape of the IV curve above 5.4 mA is caused by dissipative
movement of magnetic vortices. Switching of the detector into the resistive state around 13 mA

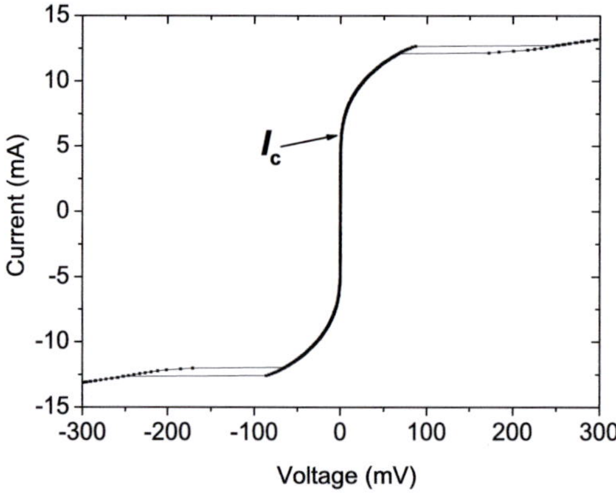

Fig. 5.9.: Current-voltage characteristic of a 30 nm thick YBCO detector (width of 4.7 μm, length of 2.5 μm)
measured in liquid nitrogen (77 K). The critical current is defined where a voltage drop of 200 μV
occurs across the bridge (I_c = 5.4 mA, see arrow). The following resistive s-shape of the IV
characteristic is determined by flux flow in the microbridge. At 13 mA the microbridge switches
to resistive state, followed by a hysteresis.

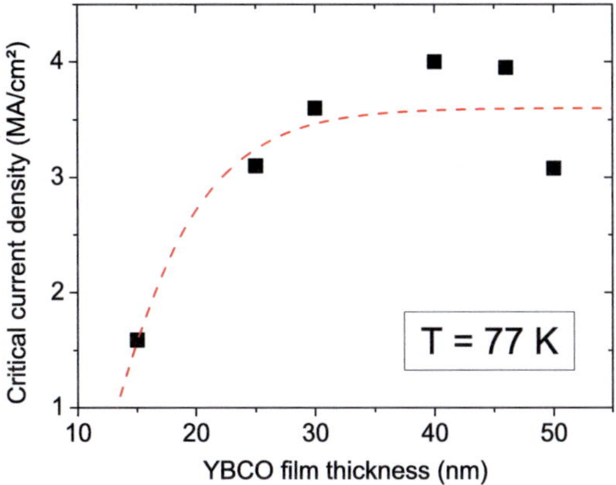

Fig. 5.10.: Dependence of the critical current density on the YBCO film thickness. For all film thicknesses high critical current densities are achieved showing good superconducting properties of our fabricated devices.

occurs via formation of a normal domain in the YBCO microbridge. By decreasing the current a hysteresis is observed. Due to noise of ± 100 µV in the measurement setup, the critical current was determined at a voltage drop across the bridge of ± 200 µV caused by the dissipative movement of the vortices. The observed critical current $I_c = 5.4$ mA (indicated by the arrow in Fig. 5.9) corresponds to a nominal critical current density of about $j_c = 3.8$ MA/cm^2.

The critical current densities in dependence on the film thickness for the YBCO microbridges is shown in Fig. 5.10. For film thicknesses above 30 nm the critical current density at 77 K is nearly constant with $j_c \approx 3.5$ MA/cm^2. For thinner film thicknesses the critical current density decreases which can be explained by the reduced critical temperature shown in Fig. 5.8.

In Fig. 5.11 the influence of the patterning process on the detecting element is studied more in detail. For this the critical temperature and the critical current density were determined in dependence of the YBCO bridge width for a constant film thickness of 30 nm. As can be seen in Fig. 5.11(a) the critical temperature decreases with the reduction of the bridge width, although some scattering of the data is observed.

In Fig. 5.11(b) the dependence of the critical current density measured at 77 K on the YBCO bridge width is shown (squares). No clear dependence of the critical current density on the bridge width was observed. The spread in j_c between 1.5 and 4.5 MA/cm^2 might be explained by the different critical temperatures shown in Fig. 5.11(a). However, also for $T = 0.9\, T_c$ (circles) no dependence of the critical current density on the bridge width is observed.

This can be explained by deviations of the nominal, designed bridge width and the effective, superconducting bridge width resulting after patterning due to damaged edges. To analyze this influence the resistivity of the patterned detectors is plotted in Fig. 5.12(a). The data shows an increase of the resistivity with decreasing bridge width confirming the assumption that the nominal bridge width is larger than the effective, superconducting bridge width.

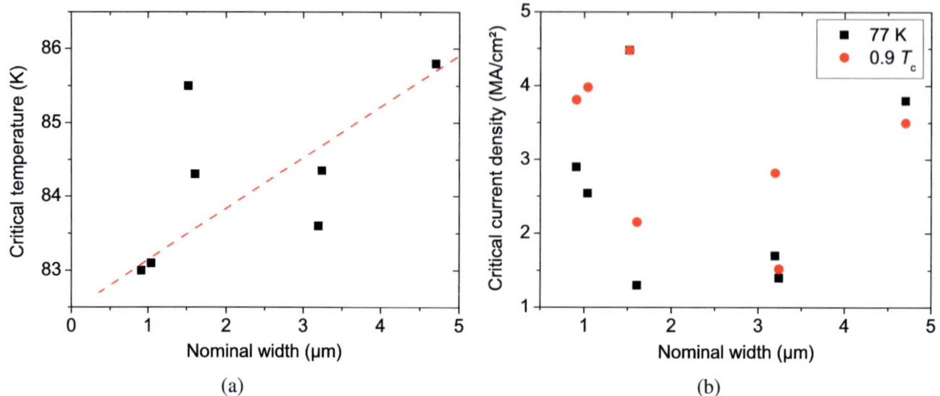

Fig. 5.11.: (a) Dependence of the critical temperature on the YBCO bridge width for the 30 nm thick patterned detectors [Raa12]. With the reduction of the bridge width the critical temperature decreases. (b) Dependence of the critical current density on the YBCO bridge width for 77 K and 0.9 T_c [Raa12]. No clear dependence of the critical current density with the YBCO bridge width is observed.

For bridge widths above 3 µm the resistivity approaches asymptotically $\approx$ 850 µΩcm, so this value is assumed as the material resistivity for our multilayer structure. By that the effective bridge width w' was calculated and is shown in dependence on the nominal bridge width w in Fig. 5.12(b). Additional influence which might be caused by the damaged surface of the YBCO layer by the wet-etching process is neglected here. The solid line in Fig. 5.12(b) shows a linear fit according to

$$w' = w - (483 \pm 124)nm. \tag{5.1}$$

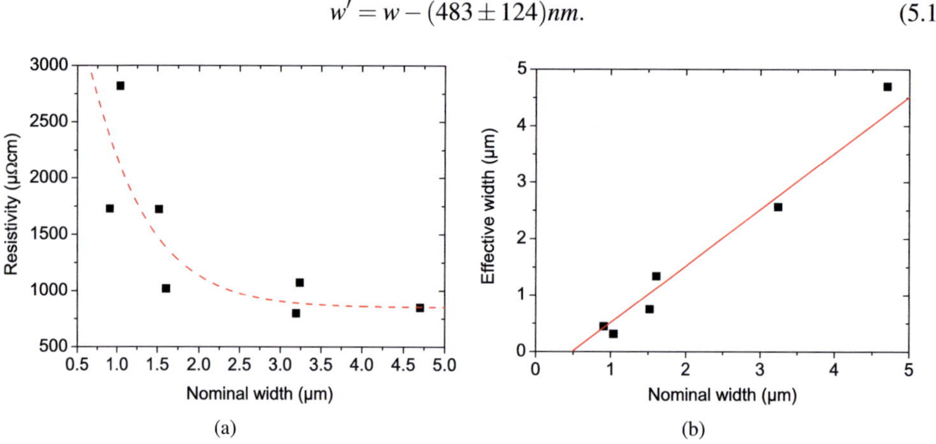

Fig. 5.12.: (a) Dependence of the normal state resistivity on the YBCO bridge width for the patterned detectors [Raa12]. With the reduction of the bridge width the resistivity increases. (b) Dependence of the effective width on the nominal bridge width [Raa12]. The solid line shows a linear fit to the data.

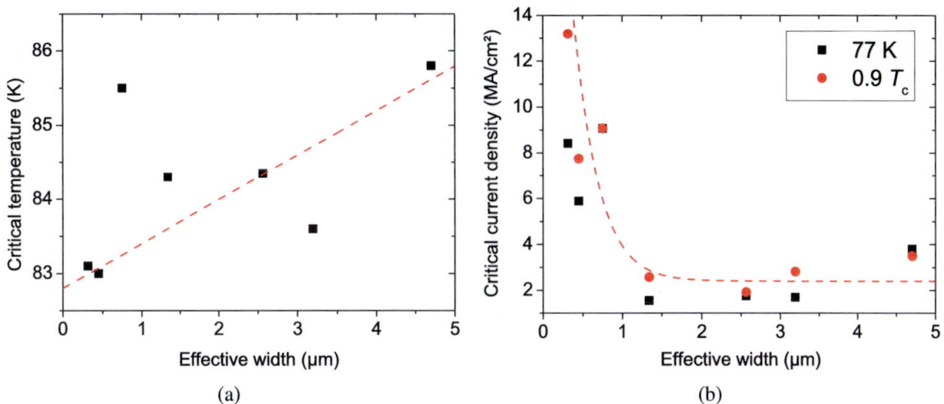

Fig. 5.13.: (a) Dependence of the critical temperature on the effective YBCO bridge width for the patterned detectors [Raa12]. With the reduction of the bridge width the critical temperature decreases. (b) Dependence of the critical current density on the effective YBCO bridge width for 77 K and 0.9 T_c [Raa12]. The critical current density increases with the reduction of the effective bridge width.

The slope of 1 shows that a constant offset between the nominal and the effective bridge width has to be taken into account. The offset amounts according to equation 5.1 to 480 nm resulting in the fact that $\approx$ 240 nm of the YBCO bridge edges are destroyed by the ion milling procedure.
Taking into account the effective YBCO bridge width the critical temperature and the critical current density are plotted in Fig. 5.13. For the critical temperature no significant deviation from Fig. 5.11(a) is observed. However, for the critical current density a clear dependence on the effective bridge width is now observed. For effective bridge widths below 1 μm the critical current density increases significantly. This might be explained by the trend towards homogeneous current distribution for the sub-μm bridges where the bridge width becomes in the order of or smaller than the effective penetration depth Λ. The effective penetration depth can be calculated according to $\Lambda = 2\lambda^2/d$, where λ is the penetration depth for thick films and $d \ll \lambda$ [154]. For our 30 nm thick YBCO samples this results with $\lambda = 130\pm10$ nm [155] in $\Lambda \approx 1.1$ μm and thus in a good agreement with the observed increase in the critical current density below 1 μm wide bridges. Another contribution might be due to changes in the pinning mechanism for the nanometer-sized bridges.

In summary, the characterisation of the superconducting properties of our patterned detectors revealed good superconducting properties with high critical temperatures ≥ 83 K and high critical current densities above 1.5 MA/cm^2 at 77 K. This makes our detectors suitable for the integration in a detecting system operating at 77 K which was demonstrated even for our ultra-thin film detectors of only 15 nm thickness in [PSH$^+$12b].

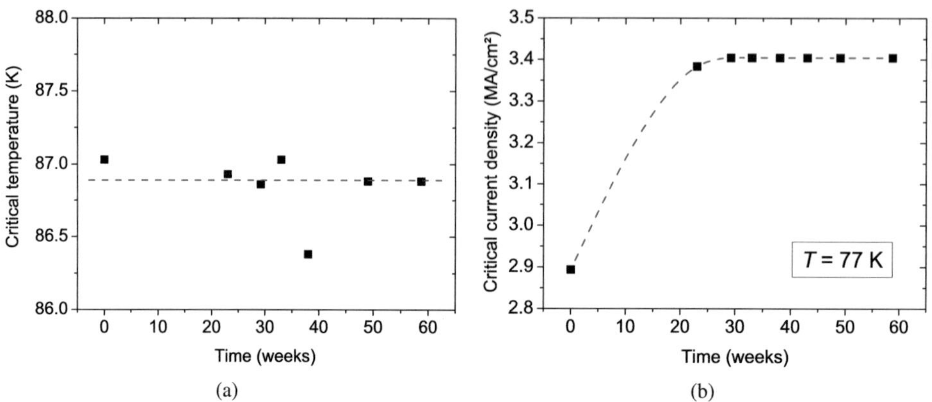

Fig. 5.14.: (a) Critical temperature of a 50 nm thick detector (width of 4.7 μm, length of 2 μm) over time. For more than one year no degradation of the critical temperature was observed. (b) Critical current density of a 50 nm thick detector (width of 4.7 μm, length of 2 μm) over time. Directly after fabrication a somewhat reduced critical current was measured. After 20 weeks a constant value was reached showing no degradation for more than one year.

5.2.2. Long-term stability

For real detector applications i.e. a permanent installation of a YBCO detection system e.g. at an electron storage ring, the long-term stability of the superconducting properties of our YBCO detectors need to be ensured. Thus, a long-term study over a period of more than one year was carried out, monitoring the critical temperature and the critical current density at 77 K for a 50 nm thick YBCO detector with a width of 4.7 μm and a length of 2 μm. Fig. 5.14(a) shows the critical temperature over time. For more than one year no degradation of the critical temperature of 86.8 K was observed.

Also the critical current density in liquid nitrogen at 77 K was monitored for 60 weeks (see Fig. 5.14(b)). For all detectors an increase of the critical current density after the initial measurement right after the fabrication was observed. This is probably explained by a relaxation and/ or reorganization of the crystalline lattice after the fabrication procedure. After this increase a constant value of 3.4 MA/cm^2 was achieved showing no degradation over a period of about 40 weeks.

In summary, the long-term study of the superconducting properties of our YBCO detectors revealed no degradation over a period of more than one year. This shows that the fabricated devices are suitable for a permanent application.

5.3. Conclusions

In section 5.1 the patterning technology for our YBCO thin film detectors was discussed. To develop a device suitable for wavelengths from the optical to the THz frequency range, a patterning process was developed to embed a micrometer or sub-micrometer sized YBCO detector in a several hundred micrometer large planar THz antenna.

The log-spiral THz antennas were designed to encompass a broad THz frequency range from 0.07 - 2 THz. The patterning process was realized by electron-beam lithography and physical as well as chemical etching procedures. The greatest challenge was the development of a patterning process for the sub-micrometer sized YBCO detector elements which was realized by a combination of Ar ion milling and wet etching in a diluted $I_2 KI$-solution. With this optimized process YBCO detector elements as short as 300 nm were successfully realized.

The DC characterization of the sub-micrometer and micrometer sized YBCO detectors was discussed in section 5.2. For all our developed detectors with YBCO film thicknesses between 15 and 50 nm we have found high-quality superconducting characteristics with critical temperatures above 83 K and critical current densities at 77 K above 1.5 MA/cm^2. Furthermore, the long-term study of the superconducting characteristics of our detectors revealed stable parameters over a period of more than one year.

This shows that the YBCO thin-film detectors developed in this work are suitable for the long-term integration in a detection system operated at liquid nitrogen temperatures.

6. YBCO detector photoresponse from optical to THz frequencies

In this chapter, the photoresponse of the developed YBCO detectors to optical and THz radiation is studied. Whereas in section 6.1 the detector response to continuous and pulsed optical radiation with photon energies above the energy gap is analyzed by frequency- and time-domain techniques, respectively, in section 6.2 the detectors are excited with continuous and pulsed THz radiation below the YBCO energy gap. For the pulsed excitations significant differences in the detector responses to over-gap and sub-gap radiation were found which are analyzed in section 6.3 and are explained in the framework of different detection mechanisms (section 6.4).

6.1. Optical frequency range

In this section the developed YBCO thin-film detectors were excited by optical radiation analyzing the photoresponse using both, frequency- and time-domain techniques. For these measurements excitations wavelengths of ≈ 800 nm were used corresponding to photon energies of 1.55 eV well above the superconducting energy gap of YBCO.

6.1.1. Frequency-domain technique

The continuous wave measurements described in this chapter allow to measure the detector response in the frequency domain. With this measurement technique, it is possible to excite the detector only with very low power levels thus ensuring quasi-equilibrium conditions of the superconducting film which are required for the employment of the 2T model as discussed in section 3.2.1.

Setup

The schematic of the experimental setup for the photoresponse measurements of the YBCO thin-film detectors to continuous optical radiation is shown in Fig. 6.1 and was set up by Dipl.-Phys. Dagmar Henrich at the IMS. The continuous radiation is generated by a commercial Toptica laser system [156], consisting of two temperature-controlled continuous-wave laser diodes with 850 nm wavelength, which are coupled into the same single-mode optical fiber. The radiation frequencies f_1 and f_2 of the two laser diodes are slightly detuned, resulting in an amplitude modulation of the output power at $\Delta f = f_2 - f_1$ with a modulation depth of the power of about 24%. During measurements, one of the diodes operates at constant conditions, i.e. bias current and operation temperature. The temperature of the second laser diode is changed resulting in a shift of its radiation frequency. The change of the laser diode temperature which corresponds to a shift of the

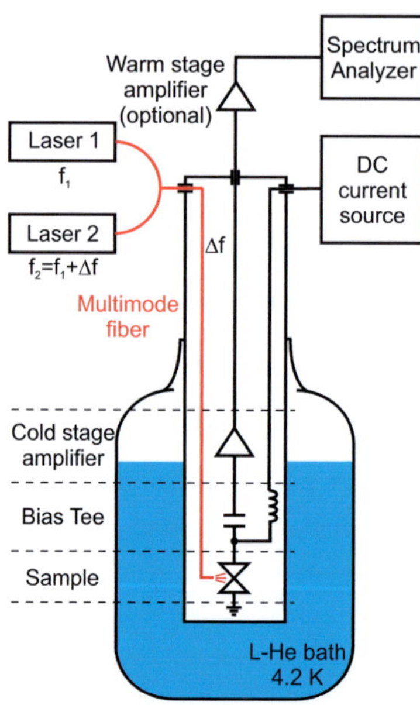

Fig. 6.1.: Schematic of the experimental setup for measurements of the YBCO photoresponse to continuous optical excitations at $\lambda = 850$ nm [RPH$^+$10].

modulation frequency from 10 MHz up to 10 GHz results in a variation of the output power less than 5%. The amplitude modulated laser power is passed into the experimental insert into a liquid helium transport dewar by an optical multi-mode fiber (105 μm core diameter) that ends at ≈ 1 mm above the sample position. Due to the numerical aperture ($NA = 0.22$) of the fiber, the beam spot at the sample position has a diameter of ≈ 1 mm, thus ensuring a homogeneous irradiation of the sample area. The modulated radiation power was measured at room temperature with a photodiode placed at the sample position. At typical measurement conditions, the density of the modulated power amounts to $\Delta P = 210$ pW/μm^2. The temperature of the sample is controlled by adjustment of the contact gas pressure in the insert and the power applied to a resistive heater placed in the vicinity of the sample. The DC bias current from a low-noise current source is applied through the DC path of the bias tee. The response of the sample to modulated radiation was amplified by a two-stage low noise amplifier (50 Ω impedance, 18 dB gain, 60 K noise temperature at 4.2 K) with low power consumption (8 mW) [157]. Then, the pre-amplified signal is led out of the dewar by a rigid stainless steel high-frequency cable and a vacuum feedthrough. The response of the sample is measured in the frequency domain by a N9020A MXA spectrum analyzer from Agilent Technologies.

Dependence of phonon escape time on film thickness

The change of the electron temperature incurred by the absorbed radiation power can be measured by the change of voltage due to a change of the temperature dependent resistance of the sample

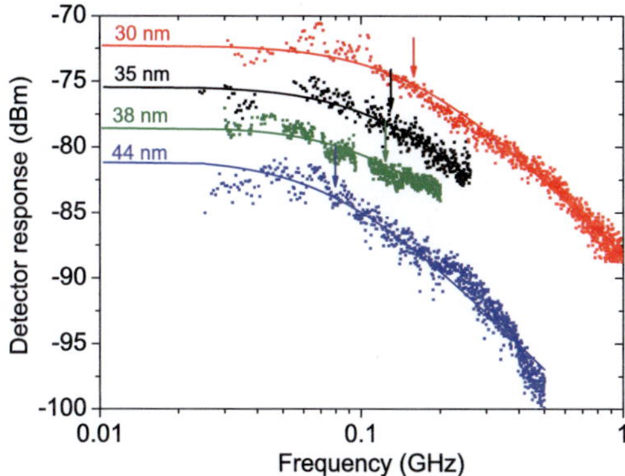

Fig. 6.2.: Dependence of voltage response ΔU on modulation frequency of amplitude-modulated optical ($\lambda = 850$ nm) power with a density of $\Delta P = 210$ pW/μm^2 for different YBCO film thicknesses as indicated in the graph. The curves in the graph are shifted in y-direction for better comparison. The solid lines are fits of the spectra by equation 6.2. The characteristic roll-off frequencies f_{es} are indicated by arrows.

according to [RPH$^+$10]

$$\Delta U = I_{bias}\frac{dR}{dT}\Delta T_e. \tag{6.1}$$

For a periodic excitation $P_{in}(t) = P_0\cos \omega t$, the time evolution of the electron temperature change ΔT_e can be calculated according to Perrin and Vanneste [105] from the 2T model (see section 3.2.1) which is discussed in detail in appendix B. The resulting equation can be simplified to a single roll-off function:

$$\Delta U(\Delta f) = \frac{\Delta U(0)}{\sqrt{1+(\Delta f/f_{es})^2}}, \tag{6.2}$$

where f_{es} is the characteristic frequency determined by the rate of energy relaxation processes and $\Delta U(0)$ is the voltage response of the sample at $f \ll f_{es}$.

Since in YBCO the heat capacity of the phonons C_p is much larger than that of the electrons C_e by a factor of 38 (section 3.2.1), the phonon and electron relaxation processes are strongly decoupled. In this case, the relaxation processes are dominated by the heat transport performed by phonons crossing the thermal boundary from the film into the substrate, so that the phonon escape time τ_{es} can be directly derived from the characteristic frequency f_{es} according to $\tau_{es} = (2\pi f_{es})^{-1}$. The much faster electron-electron and electron-phonon interaction processes occur in the frequency domain above 1 THz and can therefore not be analyzed with this measurement technique.

The measurements of the YBCO response spectra have been acquired at a constant bias current by sweeping of the modulation frequency Δf (Fig. 6.2). A plateau for the lower frequencies has been observed for all samples, followed by a decay of ΔU at higher frequencies. The experimentally measured modulation frequency spectra of the voltage response are fitted by equation 6.2, where $\Delta U(0)$ and f_{es} are the fitting parameters. The solid lines in Fig. 6.2 show these fits for the response spectra measured on YBCO detectors with different thicknesses. From these fits we extracted the

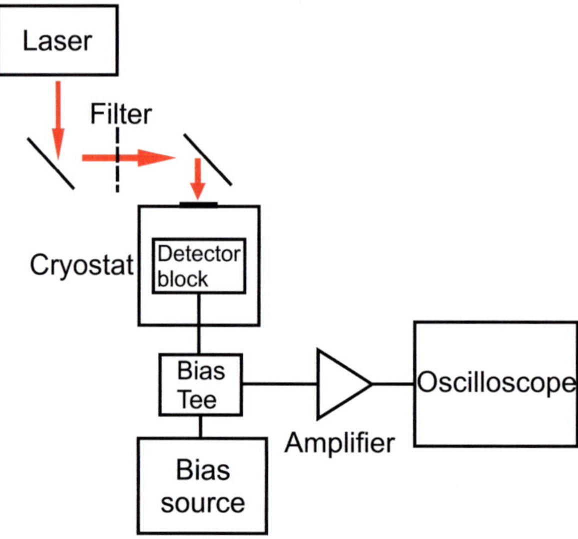

Fig. 6.3.: Schematic of the experimental setup at the DLR Berlin at 800 nm for the measurements of the detector response to optical pulsed excitations.

characteristic roll-off frequencies f_{es}, indicated by arrows in Fig. 6.2. As it is seen the f_{es} values increase with decrease of thickness of the YBCO films as expected [158]. The extracted τ_{es} values range between 1 ns and 2 ns for the 30 nm and 44 nm thick films, respectively which is in good agreement with earlier published results [158, 159].

6.1.2. Time-domain technique

In the optical frequency range a variety of femtosecond-pulsed laser sources is available. These ultra-short pulses allow to study picosecond time-domain processes in superconducting thin films, as e.g. the electron-phonon interaction in YBCO. At the same time also slower processes, as the phonon escape to the substrate on a nanosecond time scale are observable. However, due to these ultra-short laser pulses, high peak power in the kW-range is incident on the detector which can easily lead to non-equilibrium conditions in the superconducting thin film thus falsifying in particular the phonon relaxation rates. Therefore, it is important to crosscheck the achieved time constant by a low-excitation measurement technique as demonstrated in section 6.1.1.

Setup

The schematic of the experimental setup for the pulsed measurements in the time-domain located at the DLR Berlin is shown in Fig. 6.3. At 800 nm a $TiAl_2O_3$ femtosecond pulsed laser was used, which delivered 26 fs long pulses with a repetition rate of 80 MHz. The power delivered to the YBCO detecting samples was varied by a set of gray filters.

Table 6.1.: Characteristics of the YBCO thin-film detectors ($w = 4.5$ µm, $l = 2$ µm) used for the photoresponse experiments.

YBCO film thickness (nm)	Critical temperature (K)	Critical current at 77 K (mA)	Critical current density at 77 K (MA/cm^2)	Normal state resistance (Ω)
15	83.4	1.6	2.4	140
30	85.2	3.2	2.3	170
50	86.4	6.9	2.9	100

The laser light entered the continuous-flow cryostat through the optical window. A small opening in the detector housing granted free access of the laser light to the sample. The detector block was connected via a semi-rigid cable to a vacuum feed through leading to a room-temperature bias tee. The detector signal was read out via a room-temperature amplifier and a sampling oscilloscope (LeCroy SDA100G). The detector was biased with a battery-driven source.

To determine the absorbed energy in the detectors during the photoresponse experiments the maximum averaged input power was measured and amounted to 23 mW. With the 80 µm beam diameter, a pulse energy of $E = 0.69$ pJ was estimated to be delivered to the microbridge area of $9 \cdot 10^{-12}$ m^2 (width: 4.5 µm, length: 2 µm). A Gaussian beam profile was assumed and losses at the cryostat window were taken into account. To obtain the absorbed energy the absorption coefficient for films with the skin depth much less than the wavelength was used [148],

$$A = \frac{4Z_0 R_{sq}}{(Z_0 + (n+1)R_{sq})^2} \tag{6.3}$$

where Z_0 is the impedance of free space, R_{sq} the square resistance of the microbridge in the normal state, and n the refraction index of sapphire. With $Z_0 = 120\pi$ Ω, $n = 1.7$, and $R_{sq} = 320$ Ω, we found the absorption coefficient $A \approx 0.3$. The output power of the laser was varied with a set of gray filters resulting in absorbed energies per pulse from 206 to 9 fJ [PSR$^+$12].

Dependence of the detector response on bias current

At $\lambda = 800$ nm a detailed study of the detector response on different operation conditions, namely bias current, operation temperature and absorbed energy level was carried out for three YBCO microbridges with different film thicknesses (15 nm, 30 nm and 50 nm). The detailed characteristics of these three detectors are summarized in Table 6.1.

In Fig. 6.4(a) a set of typical pulses for optical excitations is displayed. The detector response of the 30 nm thick sample operated at 75 K at an absorbed energy level of 206 fJ is shown in dependence on the bias current. All pulses show a fast component in the beginning of the response corresponding to the electron heating and relaxation processes as discussed in section 3.2.1. The electron thermalization time which occurs on a sub-picosecond time scale and the electron-phonon

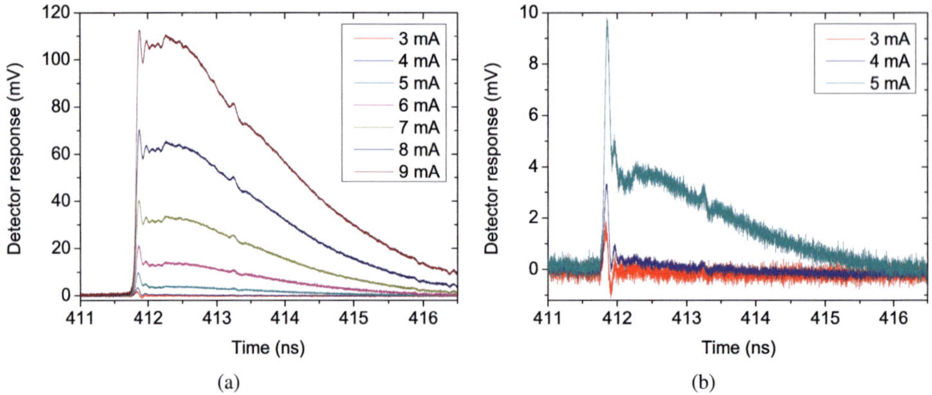

Fig. 6.4.: (a) Detector response to optical pulsed excitations ($\lambda = 800$ nm) at an absorbed energy level of 206 fJ in dependence of the bias current as indicated in the legend. The 30 nm thick sample was operated at 75 K. (b) Zoom of the detector responses at low bias currents. For very small bias currents (≈ 3 mA) only the fast component of the response is observed.

interaction causing the electron relaxation which occurs on a picosecond time scale can be clearly distinguished from the nanosecond component of the pulse corresponding to the phonon escape to the substrate. Obviously, the sub-picosecond and picosecond processes appear on a somewhat broader time scale due to the limited bandwidth of the electronic readout. Also the small peak on the nanosecond decay component is caused by the readout due to reflections in the readout chain. In Fig. 6.4(b) a zoom of the detector responses for small bias currents is displayed. For small bias currents of ≈ 3 mA only the fast component of the detector response is observed. The nanosecond component is completely absent since the YBCO thin-film bridge is only heated to temperatures slightly above the critical temperature and is already at superconducting temperat-

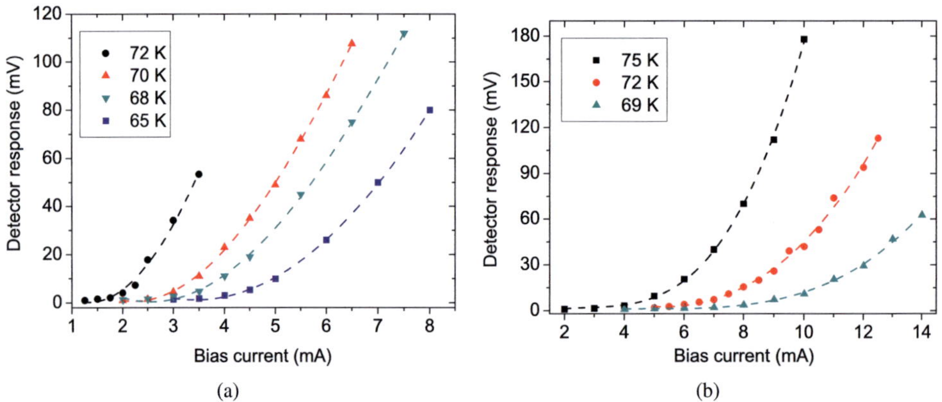

Fig. 6.5.: (a) Dependence of the 15 nm thick YBCO detector response to optical pulsed excitations ($\lambda = 800$ nm) in dependence on the bias current for different bath temperatures. (b) Dependence of the 30 nm thick YBCO detector response to optical pulsed excitations ($\lambda = 800$ nm) in dependence on the bias current for different bath temperatures.

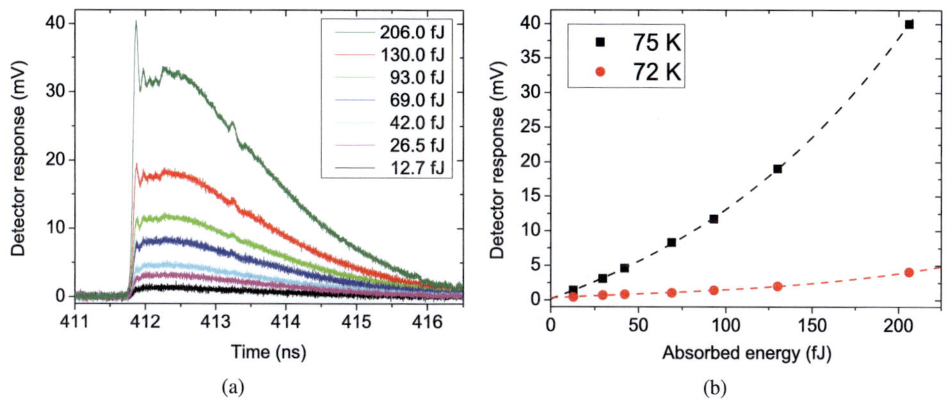

Fig. 6.6.: (a) Detector response to optical pulsed excitations ($\lambda = 800$ nm) at a bias current of 7 mA in dependence on the absorbed energy level as indicated in the legend. The 30 nm thick sample was operated at 75 K. (b) Dependence of the 30 nm thick YBCO detector response to optical pulsed excitations ($\lambda = 800$ nm) in dependence on the absorbed energy for different bath temperatures at a bias current of 7 mA.

ures when the phonon escape occurs. Exactly this mode of operation is the crucial requirement to use the autocorrelation technique for measuring the picosecond electron-phonon interaction time which is discussed below.

In Fig. 6.5 the maximum of the pulse detector response in dependence of the bias current for different bath temperatures is shown for an absorbed energy level of 206 fJ.

The detector response of the 15 nm (Fig. 6.5(a)) and 30 nm (Fig. 6.5(b)) thick sample increases non-linear with increasing bias current which is expected due to the strongly non-linear superconducting resistive transition. Furthermore, it is seen that for lower bath temperatures a larger bias current is required to get the same detector response amplitude. This is explained due to the additional required heating by the bias current to reach the superconducting transition.

Dependence of the detector response on absorbed energy

In Fig. 6.6(a) detector responses of the 30 nm thick sample biased at 7 mA are displayed in dependence on the absorbed energy level. The sample was operated at 75 K. The pulses show the same characteristics as in Fig. 6.4(a) with a fast component followed by a nanosecond decay. With increasing absorbed energy level the detector response raises as expected. In Fig. 6.6(b) the dependence of the detector amplitude on the absorbed energy is shown for the 30 nm thick sample for two different temperatures biased at 7 mA. With the decrease of the bath temperature also the detector response decreases which is explained by the increasing temperature difference between operation and critical temperature. Furthermore, a non-linear increase of the detector response is observed with increasing absorbed energy level which is again explained by the non-linear resistive transition and typically limits the linear dynamic range of a bolometric detector.

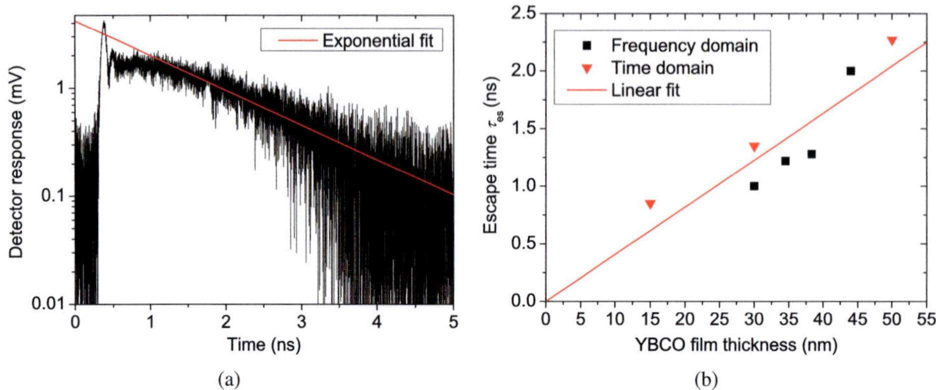

(a)   (b)

Fig. 6.7.: (a) Detector response of the 30 nm thick YBCO sample with an exponential fit to the nanosecond decay caused by phonon escape to the substrate. (b) Dependence of the phonon escape time on the YBCO film thickness extracted from time-domain and frequency-domain measurements. The solid line shows a linear fit to the data according to $\tau_{es}[\mathrm{ns}] = 0.041d[\mathrm{nm}]$.

Dependence of phonon escape time on film thickness

As discussed in the beginning of this chapter, the phonon escape time can be extracted from these pulse measurements. By fitting the exponential decay of the detector response the nanosecond relaxation component is determined. A typical fit of the 30 nm thick sample is shown in Fig. 6.7(a). The exponential fit results in a decay time of 1.35 ns. In Fig. 6.7(b) the phonon escape times for all three film thicknesses are displayed [PSR$^+$12] and compared to the extracted τ_{es} values from the frequency-domain measurements discussed in section 6.1.1. The extracted time constants from both measurement techniques show reasonable agreement and are fitted by a linear dependence on the film thickness d according to $\tau_{es}[\mathrm{ns}] = 0.041d[\mathrm{nm}]$ (see solid line in Fig. 6.7), which was found to be in good agreement with earlier published values [158, 159].

Autocorrelation technique

To determine the electron-phonon interaction time which occurs on a picosecond time scale, classical electronic readout techniques do not provide sufficient temporal resolution. To resolve time-domain processes on a picosecond and sub-picosecond time scale, electro-optical sampling or autocorrelation measurements are well-established techniques. With these techniques the electron-phonon interaction time in YBCO thin films has been determined to the single picosecond range [11, 106, 160].

In this work, the autocorrelation technique was used to measure the electron-phonon interaction time in our YBCO thin-film samples. The schematic of the autocorrelation setup from the DLR Berlin used in this work is depicted in Fig. 6.8. The fs-laser at 800 nm described above was guided to a beam splitter and subsequent mirrors. While one mirror is kept at a fixed position the other one is movable allowing to vary the delay between the two beams. The reunited beam was

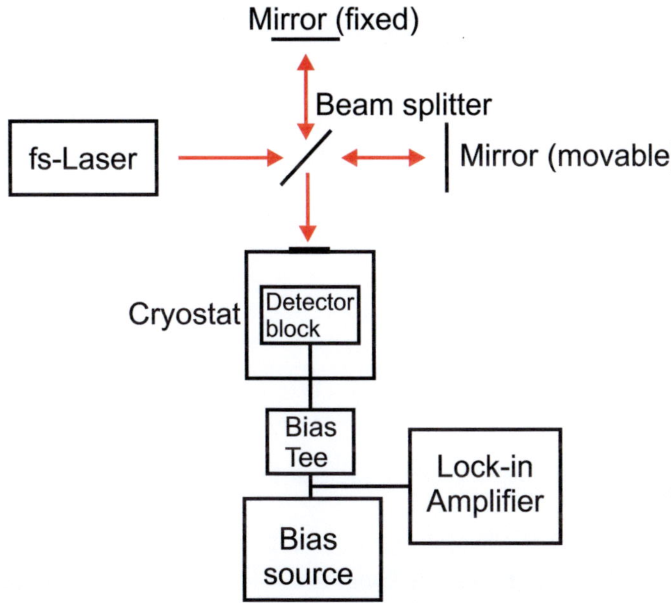

Fig. 6.8.: Schematic of the autocorrelation setup at the DLR Berlin used for the measurements of the electron-phonon interaction time in our YBCO thin-film detectors.

mechanically chopped and the autocorrelation signal was monitored at the chopping frequency with a lock-in amplifier.

Each radiation pulse with the time dependent power $P(t)$ changes the film resistance which results, for a constant bias current, in a real-time voltage transient $V[P(t)]$ across the film. The voltage across the film is measured with an integration time large compared to the reciprocal pulse repetition frequency f^{-1}. This voltage (autocorrelation signal) is given by [160]

$$U(\tau) = f \int_{0}^{f^{-1}} V\left[P(t) + P(t+\tau)\right] dt \tag{6.4}$$

where τ is the time delay between two pulses. If the response mechanism provides a nonlinearity of $V(P)$ the dependence of the autocorrelation signal on the delay time reveals the corresponding relaxation time. In the superconducting transition region the resistance of the film follows instantaneously the effective electron temperature. Thus the electron energy relaxation time is obtained while the dependence of the film resistance on the electron temperature delivers the required nonlinearity as demonstrated in Fig. 6.6(b).

The operation point of the YBCO thin-film sample was adjusted so that only the fast component of the detector response was observed on the oscilloscope (see Fig. 6.4(b)). The autocorrelation signal was then monitored with the lock-in during the movement of the mirror. The conversion from the mirror position to the corresponding temporal resolution is straightforward: 1 ps corresponds to 300 μm delay and thus motor step sizes of 150 μm. In Fig. 6.9 the measured autocorrelation

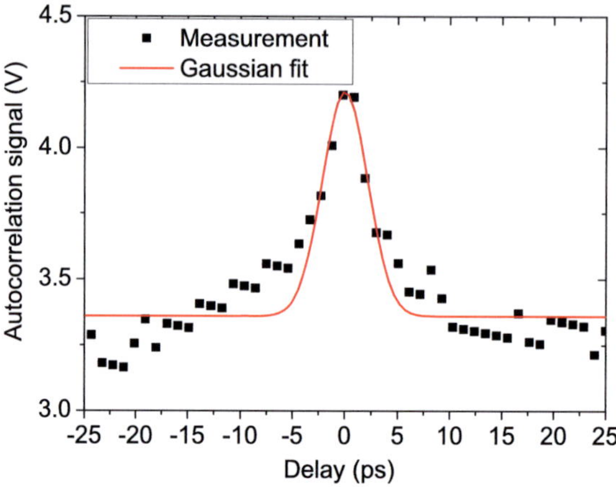

Fig. 6.9.: Measured autocorrelation signal fitted by a Gaussian function.

signal with a step size of 1 ps is displayed together with a Gaussian fit to the data. The Gaussian fit reveals a pulse width (FWHM) of 5 ps. By deconvolution of the autocorrelation signal the detector response is obtained which amounts to a pulse width of 3.9 ps. By this the ultra-fast electron-phonon interaction time in our YBCO thin-film detectors was confirmed for over-gap excitations using the autocorrelation technique showing reasonable agreement with already published values [106] which were measured with electro-optical sampling technique.

In summary, for pulsed excitations in the optical frequency range the detector response of three samples was studied in detail. The experimental data is in very good agreement with the 2T model (section 3.2.1). The fast component of the photoresponse, caused by electron thermalization and electron-phonon interaction was observed as well as the nanosecond decay caused by phonon escape to the substrate. Also the absolute values for the respective time constants show very good agreement with already published values. The dependence of the detector signal on the applied bias current and absorbed energy levels showed the expected non-linear behaviour typical for bolometric detectors.

This evidences that our developed YBCO detectors respond bolometrically to over-gap excitations in agreement with the 2T model. The fast relaxation of the electron system of only 3.9 ps confirms the motivation to use the YBCO superconductor for ultra-fast detector applications.

However, the next step is to transfer the experience from optical wavelengths to the THz frequency range and study the photoresponse of our YBCO thin-film detectors to excitations below the superconducting energy gap which is discussed in the following section 6.2.

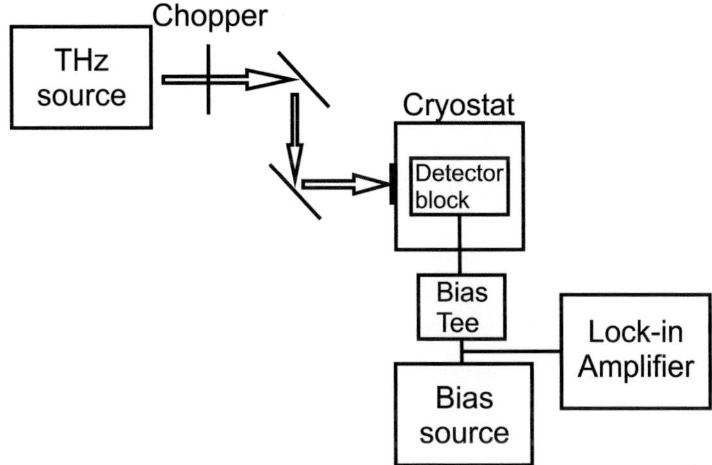

Fig. 6.10.: Schematic of the experimental setup for the continuous wave characterisation of the YBCO detectors at 0.65 THz [TRS⁺13].

6.2. THz frequency range

In this section, the developed YBCO thin-film detectors were excited by THz continuous and pulsed radiation. For the continuous sub-gap excitations a THz source at 0.65 THz was employed. The THz pulsed excitations were generated by a synchrotron source emitting frequencies mainly between 0.1 and 1 THz corresponding to photon energies below the superconducting energy gap of YBCO.

6.2.1. Monochromatic continuous wave excitations

The continuous wave (CW) measurements at 0.65 THz described in this section were employed to determine typical detector characteristics as the responsivity and the sensitivity of our YBCO thin-film detectors. Whereas the responsivity defines how large the detector voltage response is to a certain amount of absorbed power, the sensitivity defines the smallest detectable power signal.

Setup

The setup for the characterisation of our YBCO thin film detectors with continuous THz radiation is shown in Fig. 6.10 [149]. The continuous THz radiation at 0.65 THz is generated out of a 18.0556 GHz CW signal supplied by a synthesizer and transferred to 0.65 THz by a frequency multiplier chain. The output power of the THz source was determined to 110 μW using a calibrated THz powermeter from Thomas Keating Ltd [21]. The radiation is mechanically chopped and guided by two mirrors to the high-density polyethylene (HDPE) window of the cryostat. The radiation is focused by a silicon lens, embedded in a copper detector block, to the detector chip which is rear-side illuminated. The detector block is mounted to the cold-finger of the cryogen-free system. The temperature of the cryogen-free system is freely adjustable between 50 and 300 K by regulating the cryocooler compressor output power. The detector block is connected via semi-rigid cables to a vacuum feed through leading to a room-temperature bias tee. The bias tee is used to capacitively decouple the sample from the readout electronics. The detector is biased via the bias tee with a battery-driven source. All details concerning the different components of the setup and its optimization can be found in section 7.1.

The detector signal is read out via a T-adapter at the bias line with a lock-in amplifier. The lock-in amplifier measures the root mean square (rms) voltage amplitude generated by the detector at the modulation frequency. Since the generated modulation by the chopper wheel corresponds to a rectangular waveform, the peak-to-peak value of the detector signal can be reconstructed applying the following correction factor to the lock-in voltage value:

$$U_{sig} = \frac{\pi}{\sqrt{2}} U_{rms} \tag{6.5}$$

With that the optical system responsivity can be calculated according to

$$S_{opt} = \frac{U_{sig}}{P_{opt}} \tag{6.6}$$

with the optical radiation power P_{opt} emitted by the radiation source.

To determine the intrinsic responsivity of the YBCO detectors the absorbed radiation power P_{abs} in the samples has to be taken into account. For continuous wave excitation with the modulation frequency ω and the detector time constant τ where $\omega\tau \ll 1$ the frequency-dependence of the responsivity (see equation 2.8) can be neglected and S is calculated according to

$$S = \frac{U_{sig}}{P_{abs}} = \frac{(dR/dT)I_b}{G_{eff}\sqrt{1+(\omega\tau)^2}} = \frac{(dR/dT)I_b}{G_{eff}}\bigg|_{\omega\tau \ll 1} \tag{6.7}$$

with the temperature derivative of resistance dR/dT, the bias current I_b and the effective thermal conductance G_{eff} which is calculated according to equation 2.3.

To determine the absorbed optical power in the YBCO bridges at 0.65 THz the system coupling efficiency has to be taken into account. The transmission of the HDPE window of the cryocooler was determined experimentally. It amounts to 80% [STD$^+$13]. The other losses were calculated by numerical simulations. The transmission at the lens-air interface is calculated to 67% [STD$^+$13]. The Gaussian beam coupling efficiency resulted in 42%, while the polarization coupling efficiency was calculated to 50% [STD$^+$13]. This results in a coupling efficiency of 11.26%.

Finally, the coupling between the antenna and the detecting element has to be taken into account which is dependent on the impedance of the sample. To avoid uncertainties resulting from poor knowledge of the rf-impedance at 0.65 THz of our samples, the following approximation for the detector impedance Z_d will be used which originates from over-gap excitations: $Z_d = \mathrm{Re}(Z_d) + i\,\mathrm{Im}(Z_d)$, with $\mathrm{Re}(Z_d) \approx \mathrm{Im}(Z_d) \approx R_n$ [153]. Taking into account the impedance of the antenna Z_a the

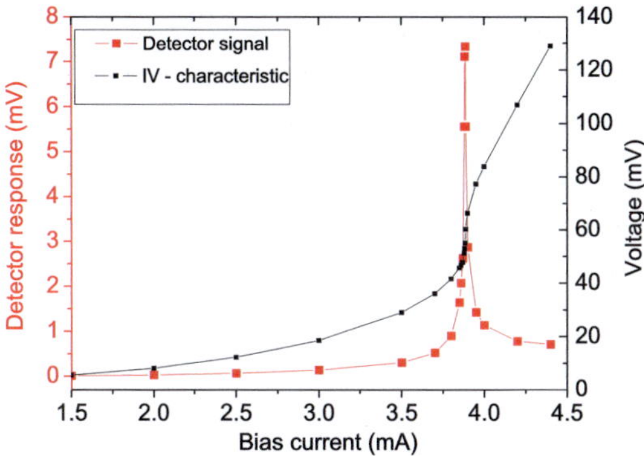

Fig. 6.11.: Dependence of the detector voltage on bias current in comparison to the detector response. The maximum response is achieved at the plateau of the IV characteristic.

coupling efficiency between antenna and detector can be calculated according to

$$\eta = 1 - |\underline{S}_{11}|^2 = 1 - \left| \frac{Z_d - Z_a^*}{Z_d + Z_a^*} \right|^2 . \tag{6.8}$$

Responsivity and *NEP*

From equation 6.7 the parameters influencing on the responsivity of the YBCO detector are derived and are discussed in this section. The bias current I_b is directly adjustable via the current source. The temperature derivative of resistance dR/dT and the effective thermal conductance G_{eff} are influenced by the bias current and the operation temperature as well as by the detector geometrical parameters.

BIAS CURRENT AND OPERATION TEMPERATURE DEPENDENCE

In Fig. 6.11 the dependence of the detector response on the bias current is shown. The detector signal increases with increasing bias current reaching the maximum at 3.8 mA before decreasing again. The maximum in the detector response coincides with the plateau in the IV characteristic shown in Fig. 6.11.

This plateau corresponds to the maximum differential resistance dU/dI of the IV characteristic. The maximum in the detector response with maximum differential resistance is clearly understood by expressing the detector responsivity (equation 6.7) by parameters only dependent on the IV characteristic of the detector according to Jones [161]:

$$S = \frac{U_{sig}}{P_{abs}} = \frac{Z - R}{2I_b R} \tag{6.9}$$

where Z is the differential resistance of the detector. The transformation from equation 6.7 to equation 6.9 is described in detail in appendix A.

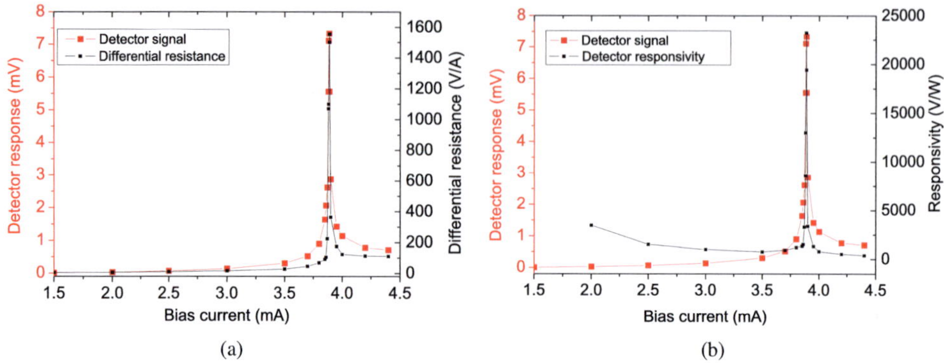

Fig. 6.12.: Dependence of detector response on bias current in comparison to the differential resistance dU/dI (a) and the detector responsivity according Jones (b) revealing good agreement between both dependencies.

The very good agreement between the dependence of the detector response and the differential resistance dU/dI on the bias current is demonstrated in Fig. 6.12(a). Thus also the calculated responsivity according to equation 6.9 shows good agreement with the measured detector response (Fig. 6.12(b)). The maximum is well represented in the calculated responsivity. However at low bias currents, when Z and R approach zero, equation 6.9 is no longer valid and the model deviates from the measurements. Furthermore, the absolute value of the responsivity reaching 23000 V/W according to equation 6.9 might be overestimated since it is defined by the differential resistance Z which is calculated from the detector bias current and voltage. However, in the plateau region of the IV characteristic, which determines the maximum dU/dI, very small changes in the current below 10 µA cause significant changes in the detector voltage. Since the used bias source only resolves currents ≥ 10 µA there might be some inaccuracies in this region. In Fig. 6.13 the

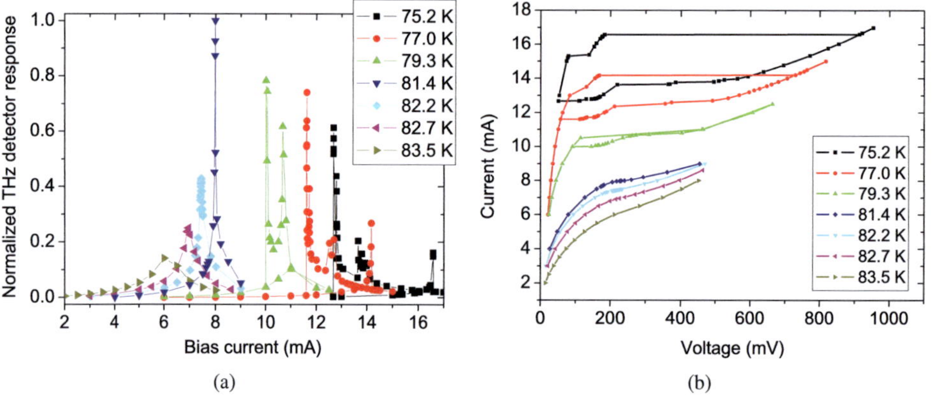

Fig. 6.13.: (a) Temperature dependence of THz detector response normalized to the maximum. The maximum detector signal was achieved for 81.4 K [TRS$^+$13]. (b) Corresponding IV curves measured at different temperatures as indicated in the legend.

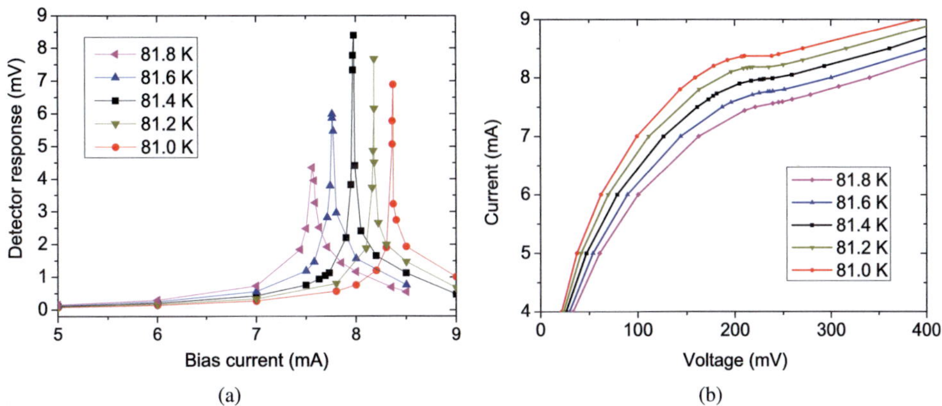

(a)   (b)

Fig. 6.14.: (a) Temperature dependence of THz detector response. At 81.4 K the maximum response is achieved. (b) Corresponding IV curves measured at different temperatures as indicated in the legend.

temperature dependence of the THz detector response and the corresponding IV characteristics for a 30 nm thick microbridge are shown. For temperatures very close to T_c the signal is small, since the detecting element is easily saturated and the differential resistance is small. For decreasing temperatures the detector signal rises reaching the maximum value for a temperature of 81.4 K. If the temperature is further reduced the detector response decreases which can be explained by the increasing temperature difference between the operation and the critical temperature, so that the temperature increase induced in the sample by the absorbed radiation is no longer enough to generate the maximum detector response.

A detailed study of the temperature dependence around the most sensitive temperature of 81.4 K is demonstrated in Fig. 6.14. It can be seen that the maximum response is achieved for the IV characteristic which still does not show any discontinuities ($T = 81.4$ K, black squares in Fig. 6.14). For this curve a stable bias point at a very high differential resistance can be adjusted.

In summary, the maximum detector responsivity is achieved at the operation temperature, where the detector shows a non-hysteretic, continuous IV characteristic. On this IV characteristic the bias current is adjusted to the plateau region where the differential resistance becomes maximum.

A further increase in the detector response might be achieved by voltage biasing allowing to adjust stable bias points also for hysteretic IV curves.

GEOMETRY DEPENDENCE

From equation 6.7 one can see that the responsivity is inversely proportional to the effective thermal conductance, whereas the latter is directly proportional to the area. Therefore, three different samples of the same thickness of 30 nm with varying detector area were analyzed. The details of the samples are summarized in Table 6.2.

Table 6.2.: Characteristics of the 30 nm thick YBCO detectors used for the responsivity study on the geometry dependence.

No.	Nominal width (μm)	Length (μm)	Critical temperature (K)	Critical current at 77 K (mA)	Critical current density at 77 K (MA/cm^2)	Normal state resistance (Ω)
1	4.7	2.0	85.7	7.5	5.3	81
2	1.85	0.67	83.4	1.5	2.7	168
3	0.9	0.3	82.9	0.9	3.3	192

For all three samples the study concerning the optimum bias current and operation temperature, as discussed above, was carried out. The result is summarized in Fig. 6.15. The optimum temperature was determined according to the IV characteristic (see Fig. 6.15(a)) showing no discontinuities and was for all detectors between 0.9 and 0.95 T_c. In Fig. 6.15(b) the differential resistance of the three samples is plotted. For samples No. 2 and 3 the differential resistance is a factor 2 larger compared to sample No. 1. This can be explained by the broader plateau region in the IV characteristic of sample No. 2 and No. 3 which is less pronounced for sample No. 1 (see Fig. 6.15(a)).

The temperature derivative of resistance is influenced by the steepness of the superconducting transition. Even for the smallest sample (No. 3) no broadening of the transition was observed compared to the microbridge (No. 1). Maximum values between 45 and 60 Ω/K were achieved for all three samples.

From the IV characteristic and the superconducting transition also the thermal conductance G of

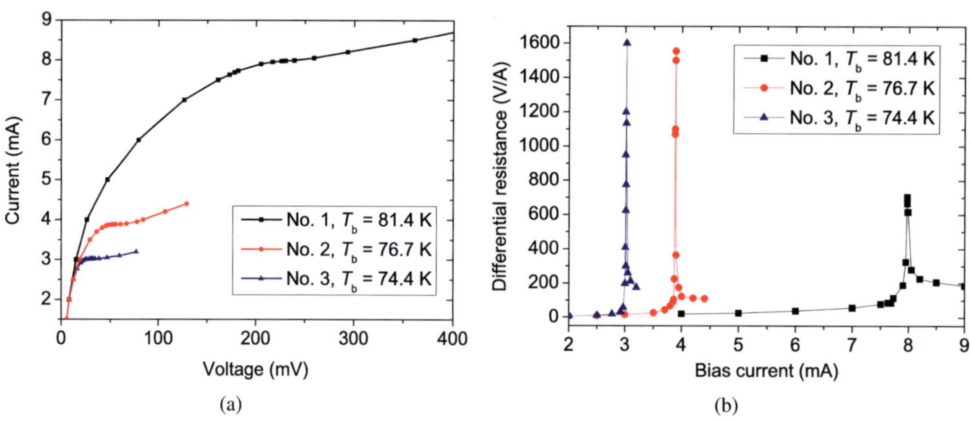

Fig. 6.15.: (a) IV characteristics of the three studied samples operated at the optimum temperature for maximum differential resistance showing no discontinuities. (b) Dependence of the differential resistance on bias current for the optimum operation temperatures for all three samples as indicated in the graph.

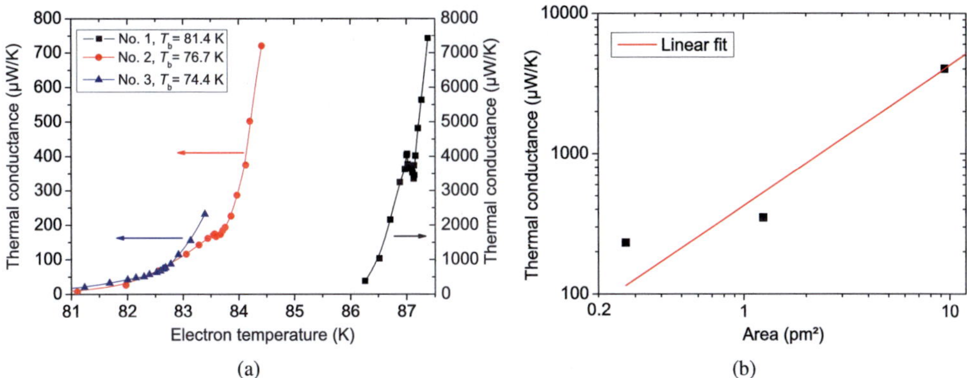

(a)

(b)

Fig. 6.16.: (a) Dependence of the thermal conductance on the YBCO detector electron temperature for the three different samples. (b) Dependence of the thermal conductance on the detector area. A linear increase of the thermal conductance with increasing detector area is observed indicated by the solid line.

the YBCO detector to the thermal bath can be calculated according to

$$G = \frac{\Delta P}{\Delta T} = \frac{U_1 I_1 - U_2 I_2}{T_{e1}(R_1) - T_{e2}(R_2)} \tag{6.10}$$

where the indices mark two neighbouring bias points on the IV characteristic defined by the voltage U, the current I and the resulting resistance R with the corresponding electron temperature T_e. Fig. 6.16(a) shows G in dependence of the YBCO electron temperature for the optimized bath temperatures for the three different samples. While sample No. 2 and 3 show thermal conductances below 1 mW/K, sample No. 1 shows a strong increase to values up to 7.5 mW/K. Using the optimum current bias point derived from Fig. 6.15(b) the resistance and thus the electron temperature of the YBCO detectors for the optimum working point is found. The comparison of the thermal conductance for the optimum working point for all three samples is shown in Fig. 6.16(b). With increasing detector area a linear increase of the thermal conductance is observed as expected (see solid line in Fig. 6.16). This confirms the motivation to fabricate sub-µm detectors for increased responsivity.

However, as demonstrated in Fig. 6.17, the maximum detector response of sample No. 2 is lower compared to the microbridge (No. 1). Only sample No. 3 can outperform the microbridge and shows a higher detector response. This is explained by two factors: First, the direct proportionality of the responsivity on the bias current (see equation 6.7). In the same way, that the thermal conductance decreases also the critical current decreased since the YBCO film thickness was kept constant at reduced width dimensions. Second, the impedance mismatch for sample No. 2 is stronger compared to sample No. 1. While this is not taken into account in the measured detector response, the responsivity calculations will compensate for that (see Table 6.3).

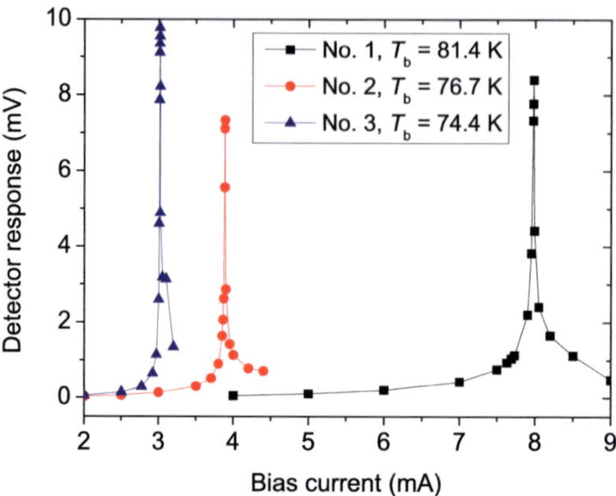

Fig. 6.17.: Dependence of the detector response on the bias current for three different samples. The sub-μm sample No. 3 shows the highest detector response.

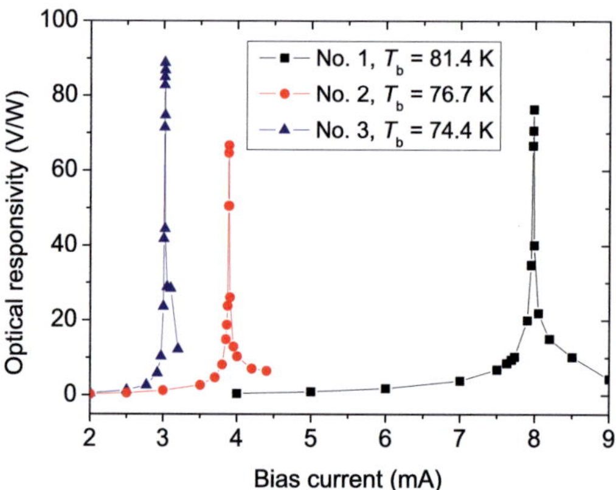

Fig. 6.18.: Dependence of the optical responsivity on bias current for three different samples. The maximum optical responsivity of 89 V/W was achieved for sample No. 3.

In summary, the lateral dimensions of the YBCO detectors influence on the responsivity as expected and as predicted by equation 6.7. With reduced lateral dimensions the thermal conductance decreases leading to an increased detector response. Possibilities to further improve the detector response are given at the end of this chapter.

ACHIEVED VALUES OF RESPONSIVITY AND *NEP*

Taken into account the emitted power of 110 μW from the 0.65 THz source the optical system responsivity can be calculated according to equation 6.6. The dependence of the optical responsivity on the bias current is shown in Fig. 6.18. The maximum responsivity was achieved for sample No. 3 and amounts to 89 V/W.

To determine the intrinsic responsivity values of the YBCO detectors the system coupling efficiency has to be taken into account as discussed in section 6.2.1. With the normal state resistance

Table 6.3.: Characterization of the 30 nm thick YBCO detectors at 0.65 THz continuous radiation. With the optical system responsivity S_{opt} and the coupling efficiency, the intrinsic responsivity S of the YBCO detectors is calculated. The *NEP* values for different modulation frequencies are given.

No.	Coupling efficiency	S_{opt} (V/W)	S (V/W)	*NEP* at 20 Hz (pWHz$^{-0.5}$)	*NEP* at 1 kHz (pWHz$^{-0.5}$)	*NEP* at 10 kHz (pWHz$^{-0.5}$)
1	8.5	76	894	2010	278	16
2	6.0	67	1117	1270	184	19
3	5.5	89	1618	309	87	12

of the samples from Table 6.2 and the antenna impedance of 65 Ω (section 5.1.1) the total coupling efficiency for the three samples can be calculated (see Table 6.3).

Taking into account the system coupling efficiency the detector responsivity values can be calculated. The found responsivity values increase with decreasing detector area as expected, resulting in a maximum responsivity for sample No. 3 of 1618 V/W.

To determine the sensitivity of our samples which is expressed by the *NEP* (see equation 2.9), the noise voltages were measured with the lock-in amplifier at the DC bias tee connector. Since the noise in the detection system is dominated by 1/f-noise the *NEP* values were determined at three different frequencies: the modulation frequency of the mechanical chopper at 20 Hz, at 1 kHz and at 10 kHz (see Table 6.3). With the reduction of the detector area the *NEP* decreases reaching a minimum value of 12 pWHz$^{-0.5}$ at 10 kHz for sample No. 3.

In this work, maximum responsivity values of 1600 V/W and minimum *NEP* values of 12 pWHz$^{-0.5}$ at 10 kHz were found for a sub-µm YBCO THz detector. In the following possible improvements of these values are discussed.

The *NEP* values are already very close to the minimum found values reported in literature (see Table 2.3). They are mainly dominated by the electronics system noise, so to further reduce these values a shielding of the detector setup should be employed.

In terms of responsivity, the values found in this work clearly outperform earlier responsivity values of substrate-supported YBCO THz detectors where maximum values of 500 V/W were achieved (see Table 2.3). Compared to free-standing YBCO bridges where responsivity values of 2900 V/W were reached (see Table 2.3) the achieved values in this work are less than a factor of 2 smaller.

To further improve the YBCO detector responsivity to CW THz radiation the detector area should be as small as possible leading to a small thermal conductance. However, the impedance matching between antenna ($\approx$ 70 Ω) and detector element has to be taken into account. Using the result of Fig. 5.12 showing that the resistivity increases with decreasing bridge width, a narrow and short detector element results in a very large detector impedance and the mismatch to the antenna impedance is rather large. Therefore, the length of the detector should be chosen as small as possible limited by technological processes to about 300 nm (see 5.1.3) whereas the minimum detector width is determined by the impedance matching.

Furthermore, the results concerning the critical temperature and the critical current density in dependence on the detector width discussed in section 5.2.1 have to be taken into account. Whereas the observed slight reduction of the critical temperature with the reduced detector bridge width should not strongly influence on the detector performance, a significant increase of the critical current density for bridge widths below 1 μm was observed. Therefore, if it is possible to reduce the resistivity for the sub-μm wide bridges by an optimized etching process and thus, to match the antenna impedance, the sub-μm wide bridges could combine several advantages: a small thermal conductance due to the small area, an adequate detector impedance and finally a high critical current density allowing to operate even a sub-μm device at moderate bias currents.

Moreover, since for continuous radiation the detector responsivity is inversely proportional to the area but not to the volume of the detector, the thickness of the YBCO layer could be increased. First of all, this guarantees a high quality of the superconducting transition even at small lateral dimensions of the detector bridge resulting in a high temperature derivative of resistance. Furthermore, the critical current of the detector is increased allowing to operate the detector at higher bias currents. Both parameters are directly proportional to the detector responsivity (see equation 6.7) thus leading to highly responsive detectors.

Finally, it is important to emphasize that the detector response to continuous THz radiation studied in this work is well explained by the bolometer model and respective equations, suggesting that the YBCO THz detector is working as a classical bolometer for continuous THz excitations. This is in agreement with earlier published work [121] where a bolometric behaviour for YBCO THz detectors was observed at and above the first maximum of the differential resistance which was studied here. Below this first maximum a vortex-based detection mechanism was reported, which could not be studied by continuous radiation in this work due to the strong impedance mismatch between antenna and detecting element in this region. However, pulsed measurements were employed to study this regime which are discussed in the following section.

6.2.2. Broadband pulsed excitations

The pulsed THz measurements were performed to analyze the detector response to sub-gap excitations in the time-domain. In earlier times this was not possible due to the lack of picosecond-pulsed high-power sources in the THz frequency range. Only the development of accelerators in recent years, made it possible to generate picosecond short pulses by coherent synchrotron radiation (CSR) in the THz frequency range (0.1 - 1 THz). This radiation was used to study the sub-gap photoresponse of our YBCO thin-film detectors.

Setup

The pulsed radiation entered the cryostat through a polyethylene window. The radiation was focused by a silicon lens, embedded in a copper detector block, to the detector chip which was rear-side illuminated. The detector block was mounted to the cold-plate of a continuous-flow cryostat and connected via semi-rigid cables to a vacuum feed through leading to a room-temperature bias tee. The detector signal was read out via a room-temperature amplifier and an oscilloscope. The detector was biased with a battery-driven source.

To determine the absorbed energy in the YBCO THz detector the coupling efficiency of the detecting system has to be taken into account [PSR$^+$12]. The THz synchrotron radiation used in this study has a repetition rate of 500 MHz. The maximum averaged output power was 3.6 mW resulting in a pulse energy of 7.2 pJ. By decreasing the electron current in the storage ring, the output power was reduced to a minimum of 30 µW (60 fJ pulse energy).

The coupling between two Gaussian modes with different beam waists was calculated according to [162]

$$K = \frac{4\omega_1^2\omega_2^2}{\left(\omega_1^2 + \omega_2^2\right)^2}. \tag{6.11}$$

The CSR beam waist was $\omega_1 = 4.8$ mm, while the equivalent waist of the lens pattern at 1 THz equaled $\omega_2 = 2.2$ mm [78]. This resulted in the coupling factor $K = 0.57$.

The reflectivity of the window and the lens surface was calculated as

$$\Gamma = \left(\frac{\sqrt{\varepsilon_r} - 1}{\sqrt{\varepsilon_r} + 1}\right)^2 \tag{6.12}$$

with the corresponding dielectric constant ε_r resulting in a total reflection loss of $\Gamma_{tot} = 0.37$.

Finally, the impedance mismatch between antenna and detecting element has to be taken into account. To analyze the differences of the YBCO photoresponse to over-gap and sub-gap excitations, the same samples as in section 6.1.2 were used. Given the frequency-independent impedance of the microbridge Z_d, the coupling efficiency between antenna and detector element was computed with standard antenna theory according to equation 6.8. Taking into account the almost real impedance of the antenna $\text{Re}(Z_a) = 65\ \Omega$ (section 5.1.1) and the normal-state resistance of the 30 nm thick sample (Table 6.1) of 170 Ω, the coupling efficiency results in $\eta \approx 0.5$.

The total coupling factor results then in $K(1-\Gamma)\eta \approx 0.2$. Taken into account the emitted CSR power, this results in coupled energies to the detector between 1440 fJ and 12 fJ [PSR$^+$12].

Since the CSR polarisation is not exactly known, losses due to the mismatch between the circular polarisation of the spiral antenna and the CSR polarisation were not taken into account. Furthermore, for sub-gap excitations the impedance of the YBCO bridge should have a much smaller real component. Its value is defined by the nature of the resistive state. However, to avoid uncertainties resulting from poor knowledge of the impedance in the resistive state, the impedance for over-gap excitation is used. Thus, the evaluated coupled energy is the upper boundary for the sub-gap case.

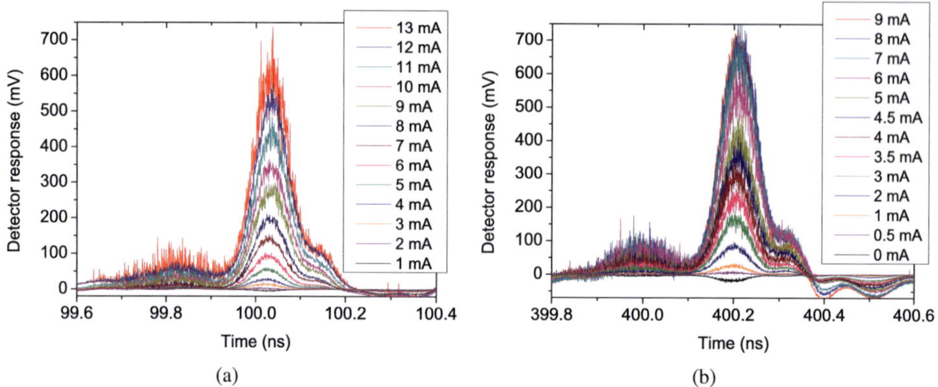

Fig. 6.19.: (a) Detector response to CSR THz pulsed excitations at an absorbed energy level of 1440 fJ in dependence of the bias current as indicated in the legend. The 30 nm thick sample was operated at 72 K. (b) Detector response to THz pulsed excitations at an absorbed energy level of 1080 fJ in dependence of the bias current as indicated in the legend. The 15 nm thick sample was operated at 65 K.

Dependence of the detector response on bias current

In Fig. 6.19 the detector response to CSR THz pulses for the 30 and 15 nm thick sample is displayed. The detector response increases with bias current showing a saturation for the highest bias currents for the 15 nm thick sample. However, even for the largest currents no change in the pulse shape is observed. For all biasing conditions no bolometric, nanosecond component is present. The detector pulse width amounts to 80 ps (FWHM) which was limited by the readout electronics. Using optimized readout electronics with a much broader bandwidth, pulse width as short as 17 ps (FWHM) were obtained (section 7.2.1).

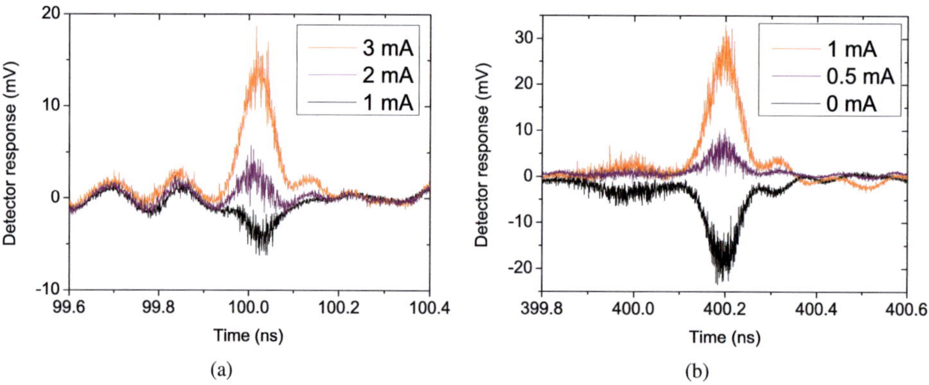

Fig. 6.20.: (a) THz detector response at $I_{bias} \leq 3$ mA for the 30 nm thick sample. The polarity of the pulse switches from positive to negative for 1 mA bias current. (b) THz detector responses at $I_{bias} \leq 1$ mA for the 15 nm thick sample. The polarity of the pulse switches from positive to negative for 0 mA bias current. A clear detector response is observed without any bias.

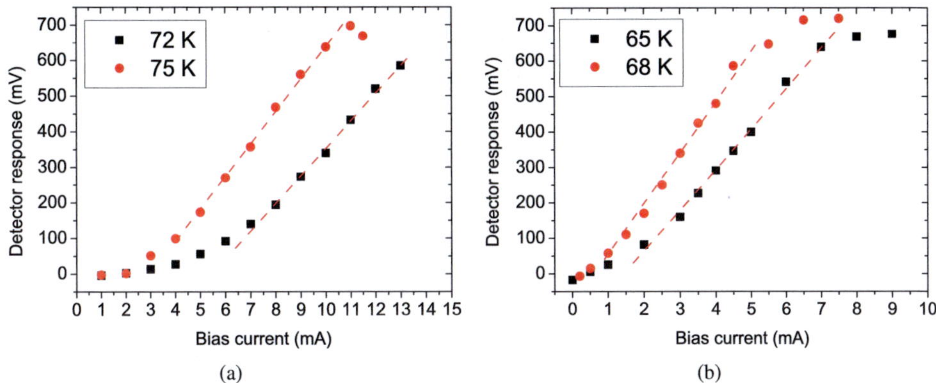

Fig. 6.21.: (a) Dependence of the 30 nm thick YBCO detector response to THz pulsed excitations in dependence on the bias current for different bath temperatures at 1440 fJ absorbed energy level. The line is to guide the eye. (b) Detector response of the 15 nm thick YBCO sample to THz pulsed excitations in dependence on the bias current for different bath temperatures at 1080 fJ absorbed energy level. The line is to guide the eye.

For the very small bias currents another striking feature was found: Fig. 6.20 shows the THz detector responses at $I_{bias} \leq 3$ mA for the 30 nm thick sample and at $I_{bias} \leq 1$ mA for the 15 nm thick sample.

With decreasing bias current a decrease of the detector response is observed until a certain threshold where the polarity of the detector pulse changes from positive to negative. While this effect happens for the 30 nm thick sample at 1 mA bias, for the 15 nm thick sample this occurs at zero-bias conditions. The zero-bias response is discussed in detail below.

In Fig. 6.21 the dependence of the detector response magnitude on the bias current for different bath temperatures is shown. For lower bath temperatures a larger bias current is required to achieve the same detector pulse amplitude. At moderate bias currents all curves show a linear dependence on the bias current as indicated by the dashed lines. For the very small bias currents and the high bias currents in the flux flow phase of the IV characteristic a deviation from this linear dependence is observed.

Dependence of the detector response on absorbed energy

In Fig. 6.22(a) a set of detector responses to different absorbed energy levels for the 30 nm thick sample is displayed. The detector response increases with increasing energy level as expected. Even for the largest absorbed energies of 1440 fJ no change in the pulse shape is observed, no bolometric nanosecond component is present.

In Fig. 6.22(b) the dependence of the detector response amplitudes on the absorbed energy levels is displayed for the 30 nm thick sample. For the whole energy range at different bias and temperature conditions a linear dependence is observed emphasized by the dashed lines in the graph. This indicates already a very broad linear dynamic range of our YBCO thin-film detectors in the THz

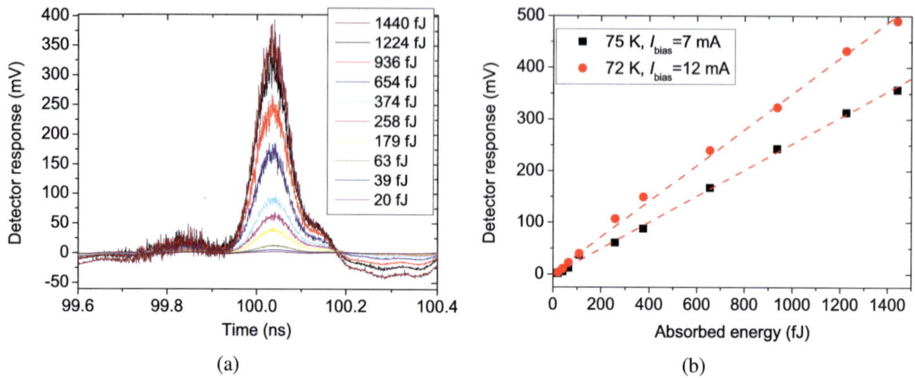

Fig. 6.22.: (a) Detector response to CSR THz pulsed excitations in dependence of absorbed energy levels as indicated in the legend. The 30 nm thick sample was operated at 75 K at a bias current of 7 mA. (b) Dependence of the detector response on the absorbed energy levels at constant bias current and bath temperatures as indicated in the legend. The dashed lines show linear fits to the data.

frequency range, contrary to the behaviour in the optical frequency range discussed above. A detailed analysis of the dynamic range of our detectors is presented below.

Zero-bias response

Zero-bias conditions have been realized by disconnecting the bias source and leaving the bias line open. For an isotropic bolometer where no thermoelectric voltage is present, no response transient can be observed without bias even for the largest available pulse energy. However, there was a clear response to THz pulses with transient magnitudes in the range of several millivolts (see Fig. 6.23(a)).

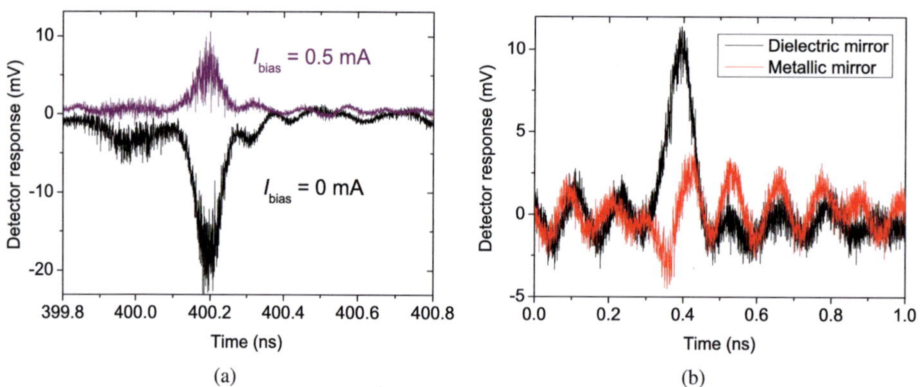

Fig. 6.23.: (a) THz detector response without any bias. (b) THz detector response of the YBCO sample at zero bias when replacing the metallic mirror with the dielectric STO mirror. The sign of the transient changes due to the change of the phase of the THz electric field [PSR+12].

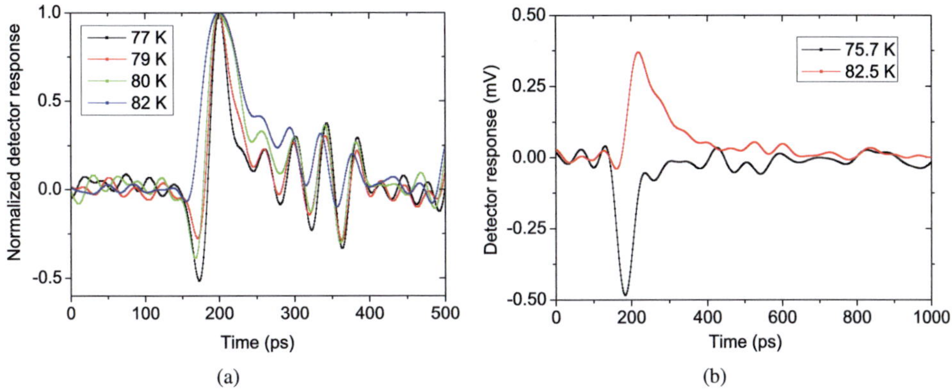

Fig. 6.24.: (a) Zero-bias detector response at different bath temperatures as indicated in the graph [Raa12]. The pulse shape changes for the highest bath temperature of 82 K. (b) Zero-bias detector response at different bath temperatures as indicated in the graph [Raa12]. The negative peak switches to a positive transient for high bath temperatures.

To prove the idea that the YBCO detector does not respond bolometrically in the THz frequency range and that the zero-bias response was caused by a mechanism driven by the electrical field of the THz wave, the metallic mirror was replaced by a dielectric mirror. By substituting the gold metallic mirror in the optical path with a dielectric mirror from strontium titanate (STO) a π-shift in the phase of the THz electric field was introduced. Fig. 6.23(b) shows the transients obtained with these two mirrors for the pulse energy of 1440 fJ. The sign of the detector transient changed, clearly indicating that the YBCO detector responds to the applied electric field of the CSR pulse [PSR+12].

The zero-bias response of the YBCO detectors was also studied in dependence on bath temperature. In Fig. 6.24 the zero-bias response of a sub-micrometer YBCO bridge ($w = 0.9$ μm and $l = 0.3$ μm) for different bath temperatures is shown. The two graphs correspond to two different fills of the electron storage ring. While for bath temperatures $T \leq 0.96\,T_c$ no significant difference in the detected pulse shape is observed, the pulse shape drastically changes for bath temperatures very close to T_c. The double polarity pulse in Fig. 6.24(a) becomes purely positive at 82 K and in Fig. 6.24(b) the polarity of the pulse switches from a negative to a positive peak. While a possible explanation for this behaviour is discussed in section 6.4, the exact mechanism occurring in the sub-μm bridge is still unclear and requires further detailed experimental as well as theoretical investigations.

Dynamic range

The dynamic range of the detector was determined by reducing the stored electron current (number of electrons in the bunch) in the storage ring, which causes a corresponding decrease of the CSR power. The operating point of the detector was in the resistive transition where the DC resistance and the differential resistance of the device were 20 and 50 Ω, correspondingly. The dependence

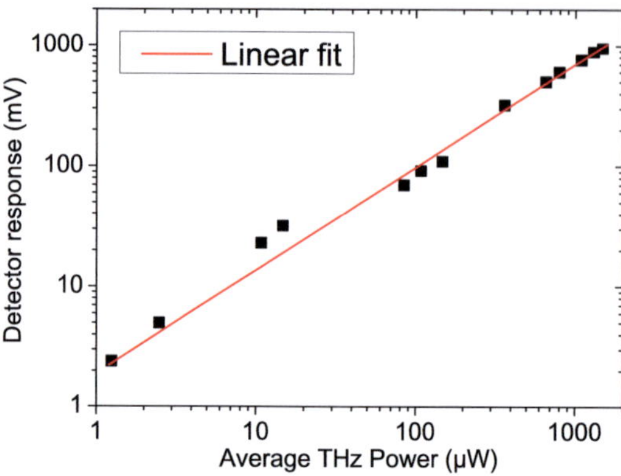

Fig. 6.25.: Detector response magnitude in dependence on the average THz CSR power [PSH$^+$11]. The solid line is a linear fit to the data.

of the detector pulse magnitude on the average THz power is displayed in Fig. 6.25. The saturation level of the amplifier was above 1 V. Over the whole measurable THz power range from about 1 to 1500 µW the detector response was proportional to the THz power, hence pointing out the dynamic range of the YBCO detector in excess of 30 dB [PSH$^+$11]. Earlier published results [58] demonstrated the saturation of a NbN HEB limiting the dynamic range to less than 15 dB.

6.3. Comparison between pulsed optical and THz detector response

In this section the reported experiments of the response transients to short optical (over-gap) and THz (sub-gap) pulsed excitations are compared as discussed in [PSR$^+$12]. It should be noted that the transients compared in this chapter at these two spectral ranges were acquired with the very same microbridges at the same operation conditions. Moreover, the energies of the pulses had comparable values. In the following the pronounced differences between the response transients at these two spectral ranges are summarized.

It was found that in the THz frequency range the time evolution of the response transient did not vary with applied bias current and had an independent duration of 80 ps (FWHM) which was limited in these experiments by the readout. Ultimate achievable time resolutions of our YBCO detectors are discussed in section 7.2.1. In Fig. 6.26 the response transients to optical and THz excitations for the 15 nm thick sample operated at 70 K for an absorbed energy level of $\approx$ 200 fJ are shown. For the optical frequency range at low bias currents only the fast picosecond component is observed. With increasing bias current the bolometric nanosecond decay appears, typical for a bolometric detector. Contrary to that, in the THz frequency range the pulse shape remained unchanged for the different biasing conditions. Even at the highest bias current of 4.5 mA which is already in the resistive part of the IV curve (see centric plot in Fig. 6.26) only a picosecond transient is observed in the THz frequency range without any bolometric component.

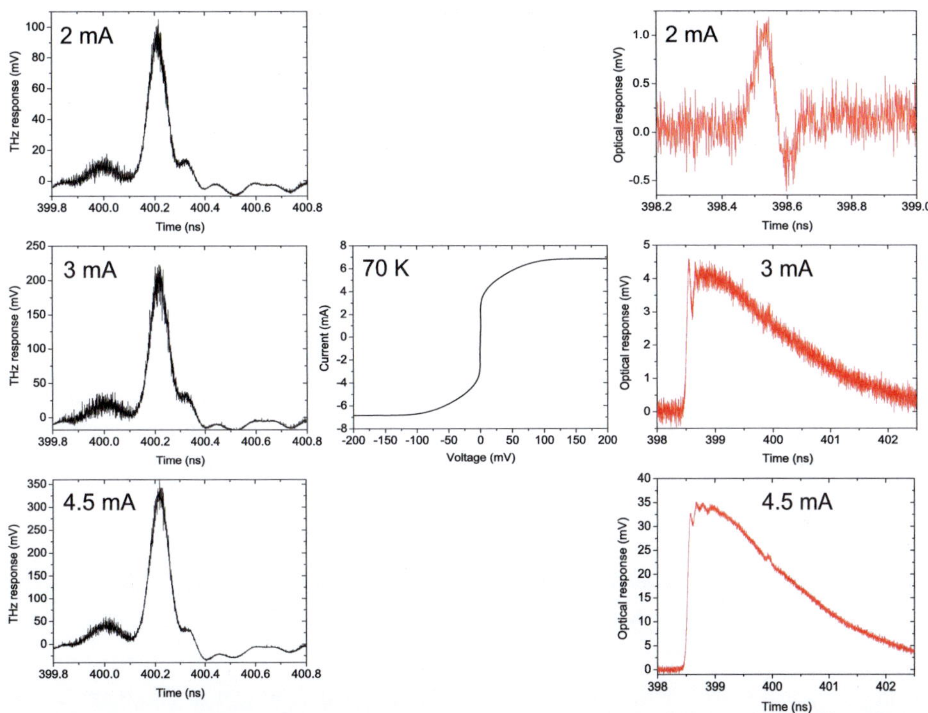

Fig. 6.26.: Response transients recorded with the 15 nm thick sample at THz (left) and optical (right) frequencies for different bias currents at an absorbed energy level of 200 fJ at 70 K. The central plot shows the corresponding IV characteristic of the sample at 70 K.

Moreover, the dependence of the THz transient magnitude on the bias current shows a deviation from the typical non-linear dependence in the optical frequency range. This is illustrated in Fig. 6.27 for the 30 nm thick sample. In the THz frequency range the transient magnitude increases linearly with increasing bias currents. Even if the absorbed energy level in the THz frequency range is large compared to the level in the optical frequency range, no non-linear behaviour is observed for the THz regime (see Fig. 6.27). While in the optical frequency range (circles) a non-linear dependence on the bias current is observed, this non-linear behavior could not be reproduced in the THz regime (squares) even for the largest absorbed energy levels in the THz frequency range of 1440 fJ.

Furthermore, the dependence of the response transients on absorbed pulse energy was studied. The pulse shape recorded in the THz frequency range was energy-independent, see left side of Fig. 6.28. In the optical range the transient changes with increasing pulse energy. At energies less than 75 fJ only the fast component is observed (Fig. 6.28 upper right). At larger energies the bolometric component with a decay time of a few nanoseconds appears (Fig. 6.28 lower right). Another significant difference between the two regimes are the transient magnitudes. At the same operation conditions and for similar pulse energies the transient magnitudes in the THz frequency

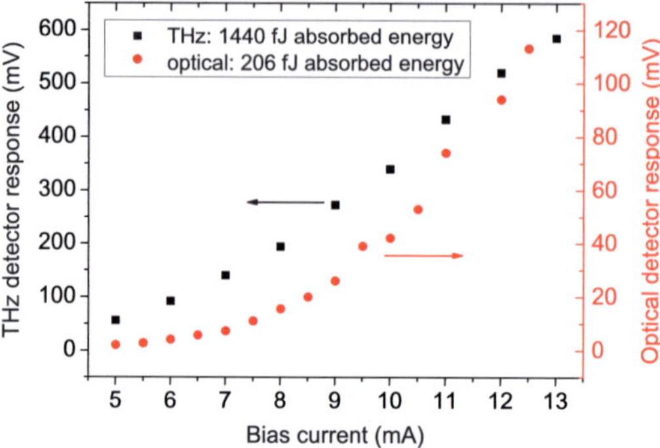

Fig. 6.27.: Dependence on bias current of the pulse magnitude for the THz (squares) and optical (circles) frequency range for the 30 nm thick sample at 72 K [PSR$^+$12]. For an absorbed energy level of 1440 fJ in the THz frequency range the detector amplitudes increase approximately linearly with increasing bias current. In the optical frequency range the non-linear bolometric increase of the detector amplitude with increasing bias current is observed for 206 fJ absorbed energy.

range are up to an order of magnitude larger than in the optical range (see Fig. 6.28).

Not only the magnitudes differ, but also the pulse-energy dependences of the transient magnitudes are different. In the THz regime the transient magnitude scales linearly with the pulse energy, while in the optical range the dependence is non-linear. The central plot in Fig. 6.28 illustrates

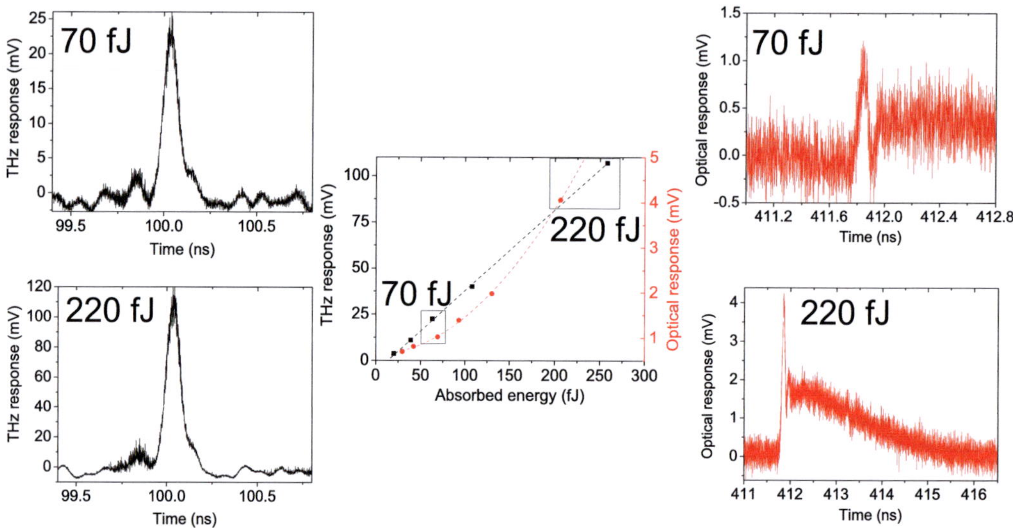

Fig. 6.28.: Response transients recorded with the 30 nm thick sample at THz (left) and optical (right) frequencies for different pulse energies indicated in the graph [PSR$^+$12]. The transients were acquired at $T = 0.85\ T_c = 72$ K and at a bias current of 12 mA (THz regime) and 7 mA (optical wavelengths). The central plot shows the pulse-energy dependencies of the transient magnitudes in these two regimes. The dashed lines emphasize the linear (THz) and non-linear (optical) dependencies.

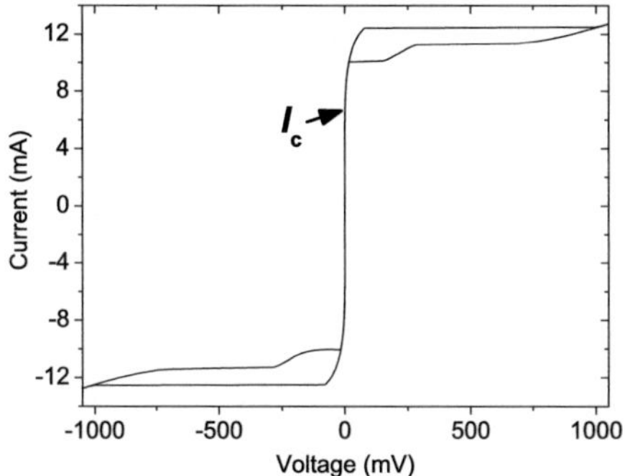

Fig. 6.29.: Current-voltage characteristic of the 30 nm thick sample acquired at 72 K with current bias.

these peculiarities for the 30 nm thick sample at $T = 72$ K. The IV characteristic, which was acquired at this temperature, is displayed in Fig. 6.29.

Finally, the most striking difference between these two regimes was found at zero-biasing conditions. Whereas at optical wavelengths no detector response was observed without any bias as expected for a bolometric detector, at THz wavelengths a clear zero-bias response was observed (see section 6.2.2).

In the following, a possible model explaining the experimental results is discussed.

6.4. The vortex-flow model

In summary, the major differences that appear in the response of YBCO bridges to short THz pulses as compared to optical excitations are [PSR$^+$12]:

(i) There is no bolometric component in the THz response. The shape of the response transient does not depend on variations of either pulse energy or bias current.

(ii) As function of both the absorbed energy and the applied bias current, the non-linear dependences of the transient magnitude in the optical range change to linear dependences in the THz range. This leads to a very broad dynamic range of the detectors in the THz frequency range.

(iii) There is a response at zero bias in the THz regime, while no zero-bias response transients have been observed in the optical range.

The two excitation regimes differ with respect to the ratio of the photon energy and the superconducting energy gap in YBCO. As discussed in section 3.1.4 the energy gap of optimally doped single-crystalline YBCO is 20 - 25 meV at zero temperature [104]. The photon energy in the optical frequency range is well above the energy gap, whereas in the THz frequency range, even for a somewhat smaller effective energy gap (see section 3.3), the photon energy is smaller than the energy gap.

The 2T model (section 3.2.1) fairly well explains the response of the YBCO bridges to optical pulses. A clear indication of the bolometric nature of the optical response is the dependence of the phonon escape time τ_{es} on the detector thickness which was demonstrated in Fig. 6.7(b).

However, in the framework of the 2T model, it is not possible to reproduce neither the magnitude nor the shape of the response transients obtained in the THz regime. One possible explanation could be that the YBCO bridge is only partly heated leading to a local bolometric response where the relaxation is dominated by electron diffusion without any significant contribution from phonon cooling. Using the known readout transfer function and the measured transient amplitudes, the length of the bridge fraction, which has to be driven to the normal state by the THz CSR was estimated. The required length is approximately 100 nm for the largest energy of CSR pulses. With the diffusion coefficient $D = 0.8$ cm^2/s (section 4.3.3) the cooling time via electron diffusion will be a few times larger than the electron-phonon interaction time at this temperature. Therefore, the slow bolometric component should be present even if the response is provided by local heating. Thus, the possibility of the local bolometric response can be ruled out.

The zero-bias response to THz CSR pulses and the polarity change, which follows the phase of the electric field (section 6.2.2), evidences that the response in the THz regime is controlled by the electric field of the CSR pulse.

To get insight in the dynamic processes during pulse detection the current which is induced in the YBCO detector by the electric field of the CSR pulse was estimated. Equating the coupled pulse energy to the Joule energy, which is dissipated within the pulse duration in the bridge with the normal resistance for the coupled energy of 1440 fJ, a current amplitude a few times larger than the used bias current is obtained. This current is definitely larger than the current at which the vortex-flow instability sets on [163]. In the current-voltage characteristic of the 30 nm thick detector (see Fig. 6.29) the onset of the vortex-flow instability is seen as a discontinuous jump in the voltage. Before the onset of the instability the DC resistance of the YBCO bridge R amounts to approximately 0.1 Ω. Assuming that this resistance is due to the movement of vortices, we estimate the corresponding vortex velocity as $2\pi d \ln(w/2d)R/(\mu_0 l) \approx 3 \cdot 10^4$ m/s, where l and w are the length and width of the 30 nm thick bridge. We also estimate the critical vortex velocity with the Larkin-Ovchinnikov (LO) theory [164] as $v_c \approx (D/\tau_{in})^{1/2}$ where D is the electron diffusion coefficient (see above). τ_{in} is the inelastic scattering time, which we associate with the electron-phonon interaction time (see section 6.1.2) resulting in $v_c \approx 4.6 \cdot 10^3$ m/s. This shows that the vortex velocity right before the onset of the instability is in the order of the critical vortex velocity. We further estimate the magnitude of the voltage transient $\approx \Phi_0 v_c/w$ (Φ_0 is the flux quantum) that corresponds to a single vortex crossing the bridge with the critical velocity and obtain the amplitude of 2.5 μV on the YBCO bridge. This is still much lower than the experimentally observed transient magnitude at zero bias (70 μV), which can be calculated taking into account the known readout transfer function and the measured transient amplitude of 3 mV at zero bias with the metallic mirror

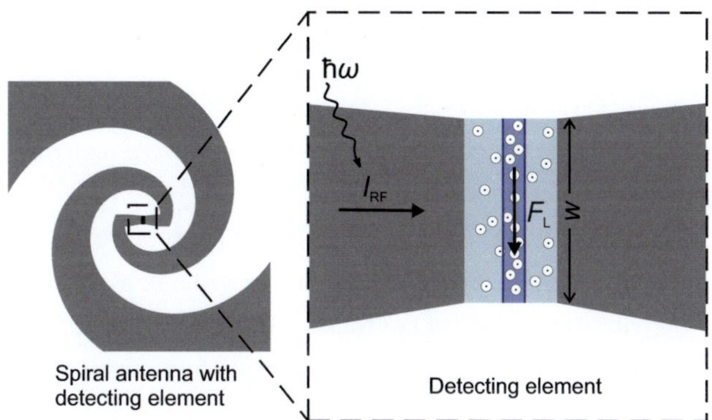

Fig. 6.30.: Schematic of vortex-flow response in YBCO THz detectors [Raa12]. The incoming radiation is absorbed by the THz antenna, generating a rf-current which causes a Lorentz Force. The Lorentz force suppresses the order parameter in the superconducting bridge which causes vortices to move giving rise to the voltage transient.

(see Fig. 6.23(b)). However, the actual vortex velocity may be higher than we have estimated. It has been shown that vortices in the LO limit change their shape and cannot be any more described as conventional equilibrium Abrikosov vortices. Additionally, several vortices in a row may cross the bridge [165]. Vortex velocities by two orders of magnitude larger than v_c were found by the numerical solution of the time-dependent Ginzburg-Landau (TDGL) equations [166] for the dimensionless parameters close to those of our bridges. On the other hand, even this velocity is not high enough to allow the vortex crossing the YBCO bridge completely within the pulse duration. Nonetheless, the possible model of vortex-assisted photoresponse discussed in the following fairly well explains our experimental data. We anticipate that the THz current pulse causes a rearrangement of the vortex lattice into a vortex line [165], which creates the channel with reduced superconducting order parameter. Vortices cross the bridge via this channel giving the voltage transient. This idea is sketched in Fig. 6.30. In the presence of the bias current healing of this channel lasts longer than τ_{in}. The actual healing time increases with the bias current. Therefore, the response magnitude in the presence of the bias current is defined by this current and the reduction of the order parameter in the channel, which was created by the THz current pulse. Without any bias the detector response can be influenced by a bath temperature variation (see section 6.2.2) which might be explained by the temperature dependence of vortex dynamics. This scenario of a vortex-flow assisted photoresponse explains qualitatively the experimental data of pulsed sub-gap excitations discussed in this work.

However, for a detailed model describing the THz response transients of the YBCO detectors, further experiments have to be carried out e.g. in magnetic field. Particular attention should be paid to the zero-bias response, since the additional contribution by the bias current can be neglected. Furthermore, the experiments should be extended to the sub-μm YBCO detectors since then the lateral dimensions might become close to the distances the vortices can travel excited by the THz pulse.

6.5. Conclusions

For the optimization of a detector technology the understanding of the detection mechanism is crucial. Therefore, in this section the photoresponse of our developed patterned samples at optical and THz wavelengths was studied in detail.

The study at optical wavelengths, i.e. for excitations above the superconducting energy gap, was discussed in section 6.1 and allowed us to compare characteristic time constants of our developed detectors with already published values in the framework of the 2T model. For this, the phonon escape times of our YBCO thin film detectors have been studied using the frequency- and time-domain technique. Both techniques revealed nanosecond time constants in dependence on the YBCO film thickness which is in good agreement with earlier published values. Furthermore, experiments using the autocorrelation technique showed an ultra-fast electron-phonon interaction time of only 3.9 ps which is again in good agreement with earlier published values and confirms the motivation to use the high-temperature superconductor YBCO for ultra-fast detector applications. Finally, a non-linear dependence of the detector signal on the applied bias current and absorbed energy levels has been found which is typical for bolometric detectors. These results showed that our detectors respond bolometrically to over-gap excitations according to the 2T model which is consistent with earlier studies of YBCO thin films at these wavelengths.

In section 6.2 this photoresponse study has been extended to THz wavelengths, i.e. to excitations below the superconducting energy gap of YBCO. Continuous wave radiation at 0.65 THz has been employed to determine the responsivity and sensitivity of our developed detectors. Maximum responsivity values of 1600 V/W were found for the nanometer sized YBCO detector as well as *NEP* values of only 12 pWHz$^{-0.5}$ at 10 kHz modulation frequency. The dependence of the responsivity on the applied bias current was well reproduced by the bolometer model suggesting that the YBCO detector is working as a classical bolometer also for the THz frequency range. This can be explained by the continuous heating of the YBCO detector inherent to the technique of CW measurements. The radiation together with the large bias current required for maximum responsivity strongly heat the electron system of the superconductor leading to a bolometric detection behaviour.

For pulsed THz radiation a completely different behaviour was observed which could not be explained in the framework of the 2T model. The YBCO detectors have been excited at the same operation conditions and with the same pulse energies as for the optical pulse measurements. However, pronounced differences were found between the optical and THz photoresponse which were analyzed in section 6.3. For the THz response no nanosecond bolometric component was found and the shape of the picosecond pulse remained unchanged for variations of the pulse energy and bias current. Furthermore, the dependence of the THz detector response on the absorbed power and bias current was linear leading to a very broad dynamic range of the THz detectors. Finally, a clear detector response has been observed at zero-bias conditions in the THz regime while no zero-bias transients were observed at optical wavelengths. This zero-bias response was excited by the THz pulse electric field rather than by its intensity which was proven by switching the polarity of the phase of the incident electric field leading to a change in the polarity of the detector pulse. In section 6.4 a possible model explaining the photoresponse at pulsed THz excitations was introduced. The idea is that the THz radiation absorbed by the planar antenna is transformed into a rf-current. Due to this rf-current a Lorentz force is generated which interacts with the vortex lattice inside the YBCO superconducting bridge causing the vortices to move. Due to the dissipative movement of the vortices the voltage detector response is generated leading to a vortex-assisted photoresponse mechanism. This work was the first detailed study for sub-gap excitations of YBCO thin-film detectors revealing clear evidence that the 2T model is not applicable for pulsed photon excitations below the superconducting energy gap.

7. Applications of direct YBCO THz detectors

The high-temperature superconductor YBCO shows ultra-fast response times and a broad dynamic range as discussed in the previous chapters. Therefore this material is a prominent candidate for fast time-domain analysis of THz pulses on a picosecond time scale over a wide power range. Since the YBCO detecting element is embedded in a broadband planar THz antenna measurements over a broad radiation spectral range are possible.
In the following chapter, the implementation of our YBCO thin-film THz detectors in a high-speed detection system is discussed. Measurements with this system at several pulsed THz sources are presented. In this work, the focus was on the monitoring and analysis of coherent synchrotron radiation emitted by electron storage rings and free-electron lasers, but also measurements at a quantum cascade laser are reported.

7.1. Detection system up to 65 GHz

To read out detector pulses on a picosecond time scale a broadband detector setup had to be developed at IMS. A schematic overview of the developed YBCO detection system for pulsed measurements is displayed in Fig. 7.1. The single components which have been build up and optimized can be divided in four groups:

1. Radiation coupling

    - quasi-optical detector block with integrated lens to couple the radiation to the detector

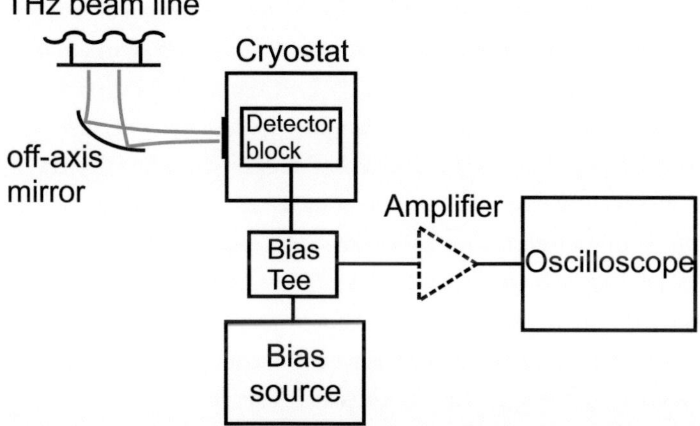

Fig. 7.1.: Schematic of the YBCO detection system for measurements of pulsed THz radiation.

2. Readout electronics

   - Rogers substrate with coplanar transmission line readout to connect the detector and readout the detector signal

   - rf-cabling up to 65 GHz for broadband readout on a picosecond time scale

   - room-temperature bias tee up to 65 GHz to bias and read out the detector

   - ultra-broadband room-temperature amplifier to read out even small detector signals

   - oscilloscope to monitor the picosecond detector pulses

3. Biasing

   - battery-driven detector bias source to avoid noise caused by mains

4. Cryogenic components

   - liquid nitrogen cryostat or cryogen-free cryocooler to cool the detectors.

In the following, these components are discussed in detail.

7.1.1. Radiation coupling

The bath cryostat and the cryocooler are equipped with two windows: one transparent for optical and infrared wavelengths (Heraeus Quarzglas GmbH) and a second one, made from highly dense polyethylene transparent for THz frequencies but opaque for optical and infrared wavelengths. The THz radiation enters the cryostat or cryocooler through the polyethylene window and is then captured by the quasi-optical detector block situated about 20 mm behind the window.

Quasi-optical detector block

The incoming THz radiation is focused to the detector chip by an extended hemispherical silicon lens ($\varepsilon_r = 11.7$) with a radius of 6 mm [TSW$^+$13]. The extension length of the lens l_{ext} determines the directivity pattern: hyperhemispherical lenses (here $l_{ext} = 1.75$ mm) are ideal for Gaussian beam coupling whereas elliptical lenses (here $l_{ext} = 2.48$ mm) have a high directivity and thus are used for plane waves [167]. For universal applicability we chose a lens extension length of 1.7 mm which results due to the sapphire substrate thickness of 0.33 mm in an effective extension lens of $l_{ext} = 2.03$ mm. The simulated radiation pattern of the antenna-lens-system for 650 GHz shows a high directivity of 31 dB as well as a high Gaussian beam coupling of 78% for a beam diameter of $2w_0 = 3.2$ mm which allows for good coupling of a focused THz beam [TSW$^+$13], [149].

The lens is embedded in a detector block made from copper (see Fig. 7.2) to ensure good thermal coupling to the cold plate. The detector block is mounted to the cold-finger or the cold-plate of the cryogen-free cryocooler or the liquid nitrogen cryostat, respectively.

The YBCO thin-film detector chip (3 mm $\times$ 3 mm) is glued to the rear-side of the silicon lens and illuminated through the backside of the substrate.

Fig. 7.2.: The YBCO detector block made from copper for good thermal contact to the cold plate. The detector chip is glued to the embedded silicon lens and rearside-illuminated [TSH$^+$12].

7.1.2. Readout electronics

For ultimate time resolution of an electronic readout system on the picosecond time scale, the effective readout bandwidth has to be evaluated and maximized. To estimate the effective readout bandwidth f_{eff}, the cut-off frequencies of all components in the readout have to be taken into account according to [9]

$$f_{eff} = \left[\sum_i f_i^{-2} \right]^{-1/2} \tag{7.1}$$

where f_i are the bandwidths of the single components.

Therefore, in this section, the cut-off frequencies of the single components in the electronic readout chain are evaluated in detail. Finally, the maximum achievable effective readout bandwidth and thus the minimum temporal resolution of the YBCO detection system is calculated.

Rogers substrate

The detector chip is connected with indium bonds to the coplanar readout on a Rogers substrate as shown in Fig. 7.3 and [TSW$^+$13].

Fig. 7.3.: The YBCO detector chip mounted to the rear-side of the silicon lens is connected via indium bonds to the Rogers substrate.

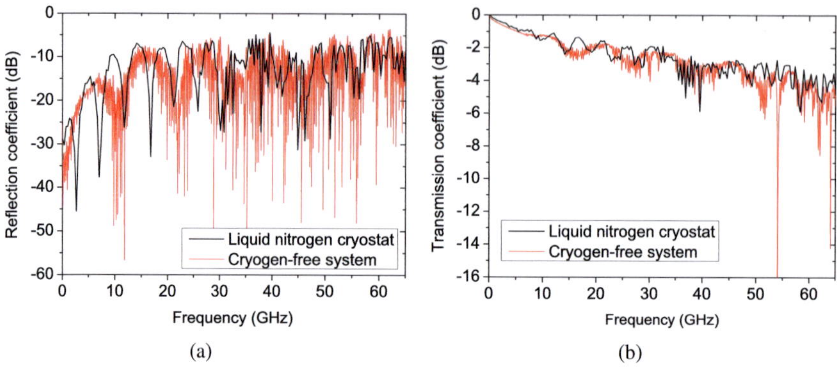

Fig. 7.4.: Measured reflection coefficient (a) and transmission coefficient (b) of the cryogenic semirigid cables in the liquid nitrogen cryostat and the cryogen-free system.

Since ultimate time-resolution on a picosecond time scale was aimed for, we decided to use high-frequency components in the V-band up to 65 GHz (IEEE-std-287 standard) to connect the Rogers substrate. A flange mount connector with glass bead from Anritsu GmbH was used. For stress relief at the interface between the coplanar line on the Rogers substrate and the connecting coaxial conductor, stress relief contacts from Anritsu GmbH were used.

The Rogers substrate embedded in the detector block including bonds and connectors was simulated with CST Microwave Studio [150]. The simulated transmission coefficient revealed a -3 dB roll-off frequency of 30 GHz [TSW+13].

Rf-cabling up to 65 GHz

For the rf-readout the cryostat and the cryocooler are equipped with three vacuum feed-throughs: two SMA feed-throughs specified from DC to 18 GHz (Huber+Suhner AG) and one V-band feed-through from DC to 65 GHz (SHF Communication Technologies AG). The rf-cabling is divided into two parts: the cryogenic cabling inside the vacuum dewars and the room-temperature cabling to connect the bias tee to an oscilloscope.

To achieve ultimate time resolution cables for the V-band specified up to 65 GHz (IEEE-std-287 standard) were used with an inner conductor diameter of 0.51 mm and a dielectric diameter of 1.67 mm. According to the cut-off formula for coaxial cables [168], which is defined as the frequency below which only the TEM mode propagates, this results in a cut-off-frequency for these cables of 67 GHz. For the cryogenic part a semirigid cable from elspec GmbH was chosen. Due to a bimetallic outer conductor consisting of stainless steel and copper, low thermal conductivity and low damping at high frequencies (7.7 dB/m at 60 GHz) can be realized at the same time. The connectors to the cable were self-assembled using V-band plugs from Anritsu GmbH. After assembly, the 30 cm long semi-rigid cables were installed in the cryogenic systems which required some bending to connect the cable to the detector block and to allow for thermalization e.g. along

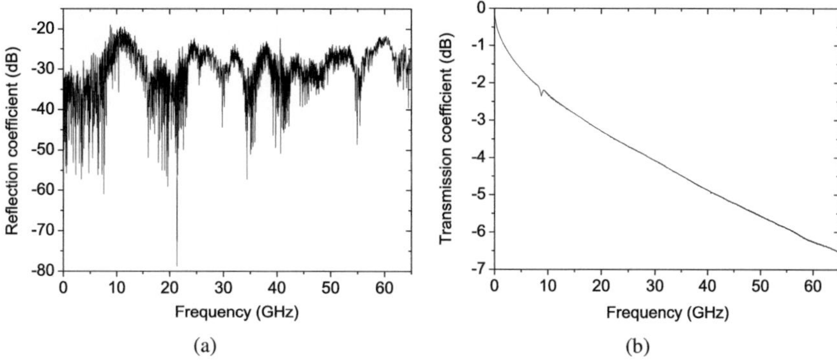

Fig. 7.5.: Measured reflection coefficient (a) and transmission coefficient (b) of the 1 m long room-temperature flexible rf-cable from Totoku electric.

the cryocooler cold-finger. In Fig. 7.4 the measurements of the reflection coefficient $|\underline{S}_{11}|$ and the transmission coefficient $|\underline{S}_{21}|$ are shown.

For both cables nearly the same characteristics were achieved with slightly worse performance for the semirigid cable in the cryogen-free system. This can be explained by the additional windings along the coldfinger of the cryocooler which were used for thermalization. The reflection parameter is for the whole frequency range for both cables below -5 dB and for most frequencies even below -9 dB which means that at least 45% up to 65% of the YBCO detector voltage pulse are transmitted. The transmission coefficient shows increasing damping with increasing frequencies as expected resulting in -4.2 dB and -5 dB at 60 GHz for the cryostat and the cryogen-free system, respectively. This means that 60% of the YBCO detector voltage pulse is transmitted.

At room-temperature flexible cables up to 67 GHz were used from Totoku electric co.,ltd. The measurements of the reflection and transmission coefficient of the 1 m long cable is shown in Fig. 7.5. At 60 GHz the transmission coefficient reveals a damping of -6.2 dB whereas the reflection coefficient is below -20 dB over the whole frequency range. This results in 45% of the voltage amplitude after the 1 m long cable. The 1.5 m long cables show a damping of -9.0 dB at 60 GHz.

Room-temperature bias tee

To bias and read out our detectors a room-temperature bias tee with a bandwidth from 50 kHz up to 65 GHz was chosen from SHF Communication Technologies AG. A maximum bias current of 400 mA at a voltage of 16 V are allowed at the bias input which is well above the requirements for detector biasing (single mA-range below 1 V, see section 5.2.1).

In Fig. 7.6 the characterization of the reflection and transmission coefficient are shown. The reflection coefficient is below -15 dB over the whole frequency range (see Fig. 7.6(a)) resulting in a reflection of 18% of the YBCO voltage pulse. The transmission coefficient shows a maximum damping of -1.2 dB (see Fig. 7.6(b)), resulting in a transmission of 90% of the voltage pulse.

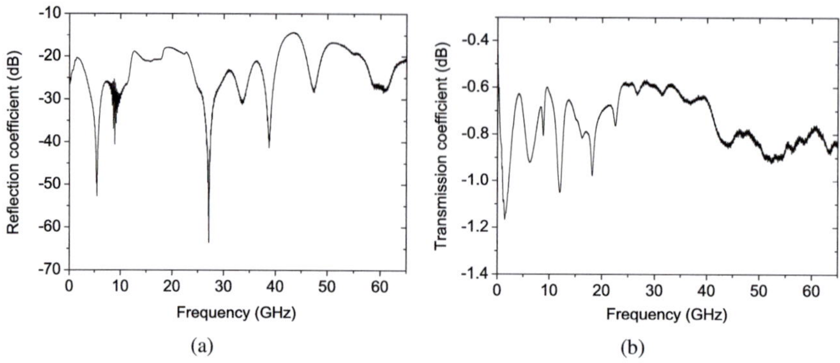

(a)   (b)

Fig. 7.6.: Measured reflection coefficient (a) and transmission coefficient (b) of the room-temperature bias tee.

In summary, if no amplifier is used in the readout chain, and the 1 m long flexible cables are used, at 60 GHz $\approx$ 10% of the generated YBCO detector voltage pulse after the detector block is still transferred to the oscilloscope. The fact that high frequencies up to 60 GHz are transferred from our YBCO thin-film detectors to the oscilloscope is emphasized by Fig. 7.7. The displayed Fourier transforms were measured with the 63 GHz real-time oscilloscope of Agilent. The cut-off of the oscilloscope (indicated by the vertical dashed line in Fig. 7.7) is clearly seen in the spectrum of our pulses which shows that high-frequencies up to 63 GHz are present in the pulses.

Room-temperature amplifier

A commercial amplifier with an ultra-broad bandwidth from 200 kHz up to 55 GHz with a specified gain of 22 dB from SHF Communication Technologies AG was used to amplify small signals. The measurements of the reflection and transmission of the amplifier is shown in Fig. 7.8. The transmission coefficient (see Fig. 7.8(b)) shows that the specified gain is fulfilled over the whole

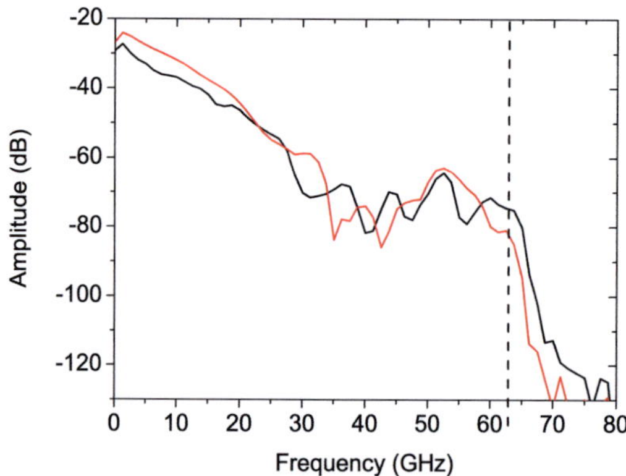

Fig. 7.7.: Fourier transforms of YBCO detector pulses recorded with the 63 GHz real-time oscilloscope from Agilent.

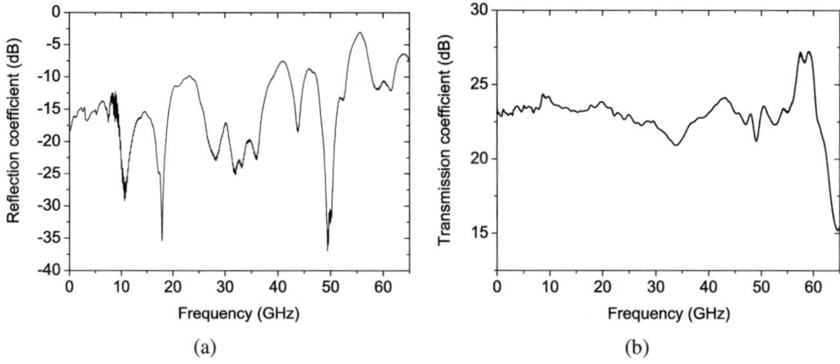

Fig. 7.8.: Measured reflection coefficient (a) transmission coefficient (b) of the room-temperature amplifier.

frequency range. The measurements of the reflection coefficient (see Fig. 7.8(a)) show a reflection below -10 dB for a wide frequency range, but also reveal the upper frequency cut-off at 55 GHz, where a reflection coefficient of -3 dB is measured.

To achieve ultimate time resolution the amplifier should not be used in the readout chain.

Oscilloscope

For the measurements described in section 7.2 different real-time oscilloscopes were used. Sampling oscilloscopes are for many experiments with ultimate time-resolution requirements not suitable. For example, at electron storage rings, due to the jitter in the arrival time of synchrotron pulses, a sampling oscilloscope is only able to measure the envelope of the arriving pulses limiting the time resolution to about 40 ps (FWHM).

To achieve ultimate time resolution of our YBCO detecting system, a close collaboration with M. Kohler and M. Stocklas from Agilent Technologies was established. They supplied us in September 2011 with a real-time oscilloscope with a bandwidth of 33 GHz. In May 2012, after the release of a new technology, we were Europe-wide the first users measuring with the 62.8 GHz real-time oscilloscope from Agilent (DSA-X 96204Q). This oscilloscope allowed us to measure for the first time, the real-time evolution of the emitted ANKA synchrotron pulses in the THz frequency range with an ultimate time resolution of 17 ps (FWHM) (see section 7.2.1).

The minimum achievable effective readout bandwidth and thus the temporal resolution of the developed YBCO detection system can now be estimated according to equation 7.1. The single component taken into account are summarized below:

The simulation of the transmission coefficient of the Rogers substrate with coplanar readout including all details of the detector block resulted in a -3 dB roll-off frequency of 30 GHz. The detector signal is then fed by the broadband semi-rigid cable ($f_{cut-off}$ = 67 GHz) to the room-temperature bias tee with a bandwidth of 65 GHz. A 1 m long flexible cable ($f_{cut-off}$ = 67 GHz) is used to feed the detector signal to the 62.8 GHz real-time oscilloscope of Agilent. Since the high

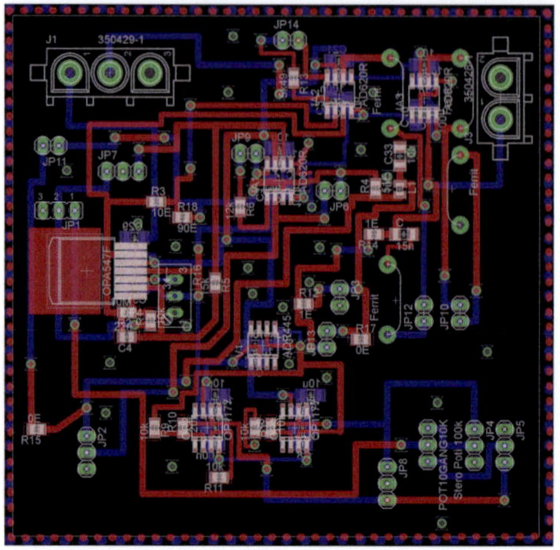

Fig. 7.9.: Board layout of the battery-driven bias source for operation of the YBCO thin-film detectors.

SNR of our detector signals allowed to readout the signal without any additional amplification, no amplifier is taken into account.

For this measurement system the effective readout bandwidth amounts to $f_{eff} \approx 22$ GHz which results in a temporal resolution (FWHM) of 15.8 ps. Measurements with this setup are discussed in section 7.2.1.

7.1.3. Biasing

For low-noise operation of superconducting detectors bias sources connected to the mains are disadvantageous. Noise caused by the mains and disturbances in the mains network lead to deterioration of the detector performance. Therefore, a battery-driven bias source was developed to bias our detectors.

Battery-driven bias source

The final board layout of the developed battery-driven bias source is displayed in Fig. 7.9.

The source was developed to operate in constant voltage as well as constant current mode, which is set by switch 1 in Fig. 7.10. To avoid any peak voltages and currents flowing over the detector element e.g. during connecting and disconnecting of cables, a short-cut knob was realized (see 2 in Fig. 7.10). The bias voltage and current were adjusted by a fine and coarse regulator (see 3 in Fig. 7.10) and was controlled by two displays (see 4 in Fig. 7.10). The source is operated with ± 12 V and supplies currents up to 50 mA (depending on the load). For a typical bias current of 5 mA the noise level of the bias source was measured at a 50 Ω load with a lock-in amplifier. At the typical modulation frequency of 20 Hz the voltage noise amounts to 300-500 nV/$\sqrt{Hz}$.

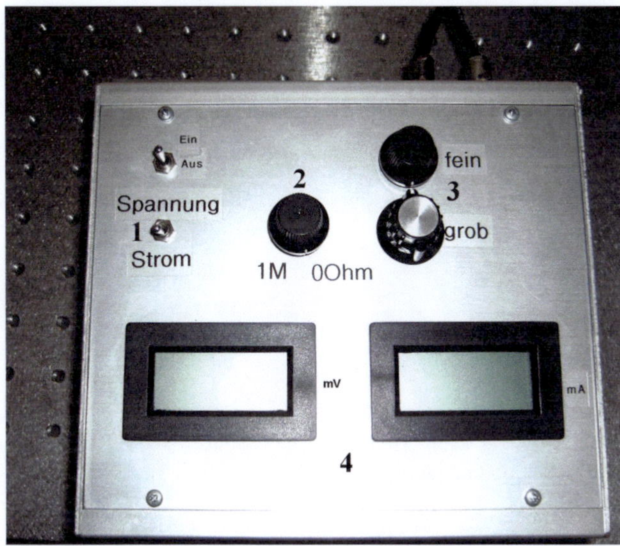

Fig. 7.10.: Battery-driven source developed to bias the YBCO detectors. (1) Switch to change between voltage and current biasing. (2) Knob to enable a short-cut of the detector bias line. (3) Coarse and fine regulator to adjust the bias current/ voltage. (4) Displays showing the bias current and voltage of the detector.

7.1.4. Cryogenic components

The high critical temperatures of our YBCO thin-film detectors allow cooling with liquid nitrogen. For cooling of our YBCO detectors two commercial systems (TransMIT GmbH) were employed: a liquid nitrogen bath cryostat and a cryogen-free cryocooler with the possibility for a fully-automated system.

Liquid nitrogen cryostat or cryogen-free cryocooler

The liquid nitrogen cryostat (embedded in a Martin-Puplett-Interferometer setup at ANKA) is shown in Fig. 7.11(a). The 850 ml liquid nitrogen dewar allows for an operation of more than 12 hours. The copper cold plate has a diameter of 96 mm with a raster of 38 holes (M4).

The long liquid nitrogen holding time and convenient filling procedure offers an easy-to-use system at the expense of a constant operation temperature of the detector of about 78 K. For absolute freedom in operation temperature and operation time a cryogen-free cryocooler was build up. The cryocooler system, shown in Fig. 7.11(b), reaches a minimum temperature of 50 K and offers a cooling power of 6.3 W at an operation temperature of 77 K. The linear compressor (Leybold Vakuum GmbH) offers in combination with water cooling of the cold finger a temperature stability $\Delta T < 0.1$ K. The cryocooler compressor power is controlled by a LabVIEW program allowing for adjusting the operation temperature freely between 50 and 300 K. The cold finger has a diameter of 38 mm, where the quasioptical detector block is attached to.

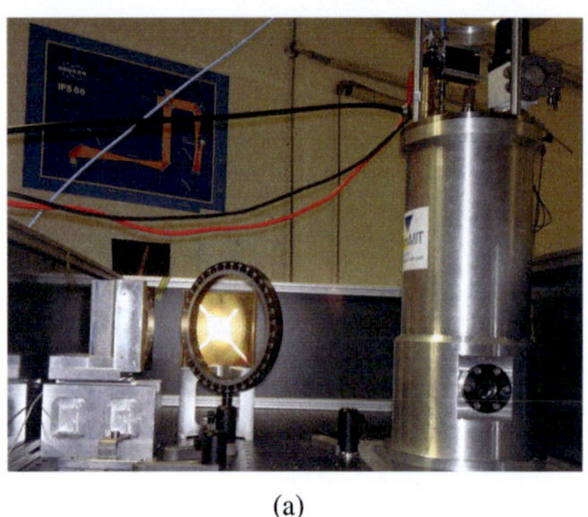
(a)

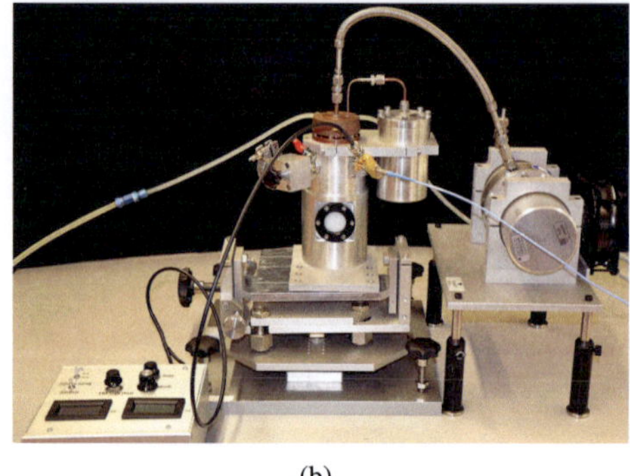
(b)

Fig. 7.11.: (a) Liquid nitrogen cryostat embedded in a Martin-Puplett-Interferometer setup at ANKA. (b) Cryogen-free cryocooler setup [TSH+12]: The compressor is connected via a flexible line to the cold-finger in the cryocooler vacuum dewar.

7.2. Measurements at pulsed THz sources

In this section the employment of the developed YBCO THz detection setup at several pulsed THz sources is discussed. One main application of the YBCO detection system is the analysis of coherent synchrotron radiation emitted from electron storage rings (section 7.2.1) and free-electron lasers (section 7.2.2). But also pulsed THz radiation generated by quantum cascade lasers is analyzed and discussed in section 7.2.3.

7.2.1. Electron Storage Rings

An electron storage ring is a particle accelerator dedicated to the emission of synchrotron radiation, providing a high brilliant beam of photons covering an extremely broad spectral range. Synchrotron radiation is emitted when electrons moving at relativistic speeds interact with a magnetic field, and covers a spectral range extending from the hard X-ray to the THz range [169].

One major application of the emitted radiation is the calibration of radiation sources and detectors [170]. Electron storage rings are used as primary sources standards for radiometry at several national metrology institutes. Further applications are for example in spectroscopy where Fourier-transform infrared synchrotron microscopy is employed to analyze chemical sample compositions [171].

Besides the fact that the spectrum of electron storage rings covers a wide spectral range and is calculable from fundamental electrodynamics relations, electron storage rings have some other interesting properties:

The electrons in a storage ring do not form a continuous stream but are grouped in so-called bunches. This is a direct consequence of the radio frequency (rf) system used to transfer power to the beam leading to a pulsed THz emission which is very interesting for time-domain studies.

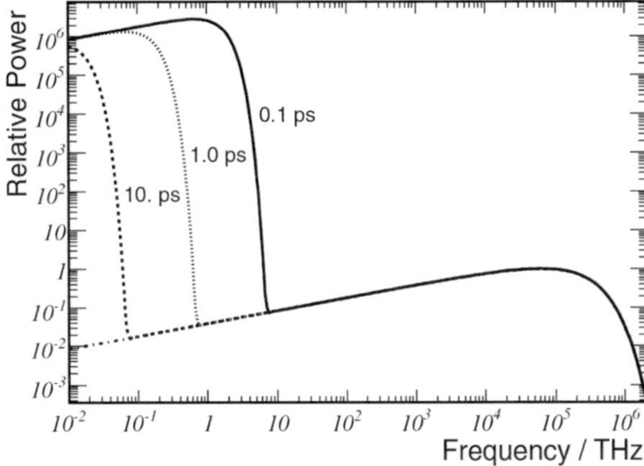

Fig. 7.12.: Spectrum of synchrotron radiation showing the amplification due to coherent emission in the low frequency region. The spectrum is shown for three different electron bunch lengths (root mean square). It is clearly visible how the coherent part of the spectrum broadens for shorter bunch lengths [1].

The synchrotron radiation power emitted by N_e electrons in a bunch can be calculated using the following equation [4]:

$$P_{total} = N_e P_e \left(1 + N_e f_\lambda\right), \tag{7.2}$$

where P_e is the (incoherent) radiation power per electron and f_λ is a form factor that is given by the length and shape of the bunch. In the case of a Gaussian density distribution of the root mean square bunch length σ the form factor is simply given by $f_\lambda = \exp\left[-(2\pi\sigma/\lambda)^2\right]$ [4]. This relation states that for short bunches and long wavelengths, the second part of equation 7.2 becomes dominant and the emitted power increases drastically: coherent synchrotron radiation is emitted i.e. the electromagnetic waves superimpose at equal phase. The enhancement of the coherent compared to the incoherent power increases with the number of involved electrons. In the case where the bunch form factor f_λ is small, the huge number of involved electrons could still result in a large radiation enhancement, given by the product $N_e f_\lambda$ [4].

Coherent synchrotron radiation (CSR) from relativistic electron bunches has been intensively investigated because of its potential ultra-high power in the THz region [172]. The first observation of CSR was achieved more than 20 years ago using 2.5 mm long electron bunches corresponding to a bunch duration of 8.3 ps produced by a linear accelerator in Japan [173]. Since then, CSR was produced exclusively by linear accelerators. However nowadays, also at electron storage rings, THz CSR is produced, using different operation regimes.

One possibility is to employ dedicated magnetic lattices, so-called low-alpha optics [4–6], to form sub-millimeter bunches leading to the emission of ultra-short THz pulses with pulse lengths in the single picosecond range [4, 7, 8]. Since the wavelengths of the emitted light are then in the range of the electron bunch length or longer, the electrons in the bunch emit coherently, which results in the quadratic scaling of the radiant power with the number of electrons in a bunch as demonstrated in equation 7.2. This leads to an enhancement of the emitted power by many orders of magnitude for those wavelengths which is demonstrated in Fig. 7.12 [1]. One can further see from Fig. 7.12 that the shape of the emitted spectrum can be varied by the adjustment of the bunch duration.

In this work, we studied with the developed YBCO detection system THz CSR produced from short bunches with single-picosecond pulse durations using the low-alpha optics at the electron storage rings ANKA (ANgstroem Quelle KArlsruhe of the Karlsruhe Institute of Technology) and MLS (Metrology Light Source of Physikalische Bundesanstalt Berlin) in Germany.

However, CSR is not only emitted from short bunches as discussed above. Two other ways of producing THz CSR were studied in this work at UVSOR-II (Ultraviolet Synchrotron Radiation Facility) in Japan employing several hundred picoseconds long bunches. One was the increasing of the electron beam current which can also lead to coherent emission due to the product $N_e f_\lambda$ in equation 7.2 as discussed above. Another technique was the so-called laser-induced CSR where longitudinal microstructures of radiation wavelength scale are produced in the electron bunch by laser interaction. The laser bunch slicing is one technique for creating sub-millimeter dip structure on electron bunches using femtosecond laser pulses [174] which interact with a small part of the electron bunch. This leads to a broadband THz emission at downstream bending magnets. At several storage rings, intense and broadband THz CSR has been successfully produced by this technique [175, 176]. Recently, another technique was demonstrated leading to narrowband THz emission, when a sinusoidal modulation of the electrons has been induced by a long laser pulse containing a sinusoidal amplitude modulation. In particular, at UVSOR-II in Japan tunable and narrowband THz CSR has been successfully produced using amplitude-modulated laser pulses [177–179].

The most common detector technology in use at electron storage rings to study the ultra-fast pulses in the time domain, is the streak camera [180]. With a time resolution down to 200 fs [181] these devices allow for real-time measurements of the temporal evolution of the pulses. However, with a cut-off wavelength in the near-infrared [181] these detectors are not suitable for the study of the coherent part of the synchrotron radiation in the THz frequency range. Furthermore, the measurement technique is indirect as the temporal pulse distribution is translated into a spatial distribution on a fluorescence screen. Thus, the signal from a single light pulse is very weak, so averaging over many pulses (or many revolutions of a single pulse) is required.

In the following it is demonstrated that the YBCO detection system overcomes these limitations resulting in successful THz CSR measurements at different electron storage rings.

ANKA - Angstroem Quelle Karlsruhe

ANKA is the electron storage ring of the Karlsruhe Institute of Technology. The ANKA electron storage ring has been in operation since September 2000. It has a circumference of 110.4 m and can be operated in the energy range from 0.5 to 2.5 GeV.

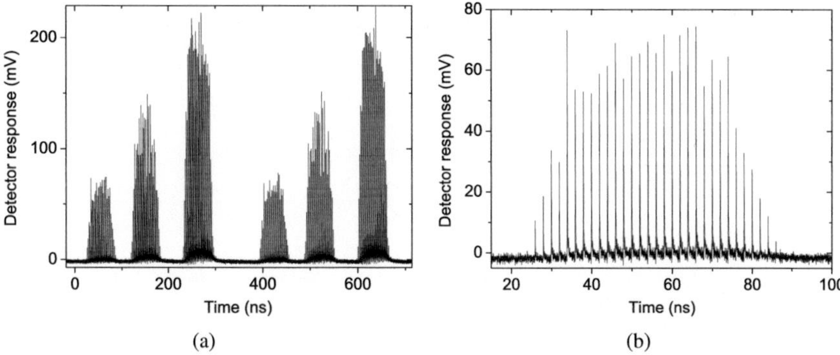

Fig. 7.13.: (a) Typical multi-bunch filling pattern of the ANKA storage ring with three trains separated by 20 ns (50 MHz) measured with a 15 nm thick YBCO detector [PSH+12b]. In (b) one train with 33 bunches is depicted in detail [PSH+12b].

THz CSR is produced at ANKA using low-alpha optics and is detected at the ANKA-IR1 beamline [169]. The synchrotron radiation which reaches the experimental hutch covers the wavelength range from 500 nm to 2.5 mm (20000 to 4 cm^{-1}, 600 THz to 120 GHz).

RESOLUTION OF THE FILLING PATTERN

The ANKA storage ring can accommodate up to 184 bunches. It can be operated with single as well as with multiple bunches simultaneously stored in the ring. The typical multi-bunch filling pattern is shown in Fig. 7.13(a). The typical filling pattern of the ANKA storage ring consists of 3 trains separated by a 20 ns gap. Each train consists of about 33 - 36 bunches. The distance between two adjacent bunches is 2 ns corresponding to the 500 MHz of the rf-system which are clearly resolved by the YBCO detectors as shown in Fig. 7.13(b).

REAL-TIME RESOLUTION OF CSR THz PULSES

The temporal resolution of the YBCO detection system was mainly limited by the real-time oscilloscope in the setup. As discussed in section 7.1.2 a 63 GHz real-time oscilloscope could be used which allowed to measure the real-time evolution of the synchrotron pulses for the first time as shown in [TSH+12]. For these measurements, the ANKA ring was operated with low-alpha mode at a beam energy of 1.3 GeV filled with a single bunch. The synchrotron frequency was set to 4.49 kHz and the voltage per cavity was chosen to 150 kV (0.6 MV total accelerating voltage). For these settings the rms bunch length σ_z is a function of bunch current as a result of turbulent bunch lengthening. It scales according to [182]

$$\sigma_z = k \cdot I_{bunch}^{3/7}, \tag{7.3}$$

with an experimentally obtained $k = 10.3$ for ANKA [180]. The bunch current I_{bunch} for the pulses shown in Fig. 7.14(a) was 0.8 mA. The rms bunch length is determined by equation 7.3

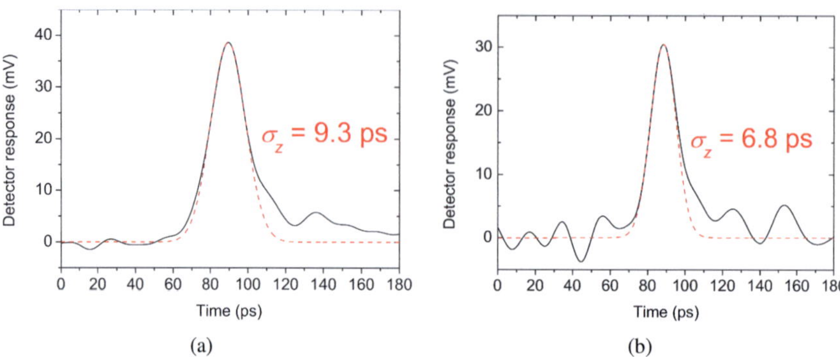

Fig. 7.14.: (a) Averaged YBCO detector response (solid line) of 20 single shots [TSH+12]. The rms pulse length was determined by a Gaussian fit (dashed line) to 9.3 ps. (b) Single shot of the YBCO detector system [TSH+12]. Pulse lengths as short as 6.8 ps were recorded.

to σ_z = 9.4 ps which corresponds to a pulse width of 22 ps (FWHM). The average of 20 single shots recorded with the YBCO detection system is shown by the solid line in Fig. 7.14(a). The Gaussian fit (dashed line) to this data revealed a rms bunch length of 9.3 ps which is in very good agreement with the bunch length according to equation 7.3. This shows that the YBCO detection system allows for the direct analysis of the emitted picosecond pulses in the THz frequency range. Furthermore, it is worth mentioning that the YBCO system is not limited to averaged measurements. Due to high signal-to-noise ratios above 10 it is possible to analyze single shots. In Fig. 7.14(b) a single shot for the ANKA settings stated above is displayed. While from 0 to 60 ps the combined noise of the detecting system and the ANKA source is seen, the detected CSR is clearly visible at about 90 ps. The bunch length was as short as 6.8 ps which is a record value achieved for the first time in the real-time monitoring of a CSR THz pulse. The falling edge is influenced by reflections which can be explained by imperfect matching of the detector impedance to the 50 Ω readout impedance. This leads to a slightly enlarged pulse width (FWHM) of 17 ps. The possibility to resolve the temporal evolution of the emitted THz pulses in single shots opens routes for the study of turn-by-turn bunch-length monitoring on a very short timescale.

BALANCED DETECTION

The high intensity of the CSR THz radiation is advantageous for Fourier Transform Infrared (FTIR) measurements, however, the signal-to-noise ratio is deteriorated by intensity fluctuations and aperiodic noise. These fluctuations are caused by the interaction of the electrons with their own magnetic field (wakefield) causing a microstructure in the electron bunch, in turn resulting in bursts of THz radiation [183]. To correct for the CSR instabilities a detection system based on two THz detectors is required to record with the first one the reference signal while the second one is used for the FTIR spectroscopy of a probe. The proof of this concept of balanced detection was demonstrated using slow composite bolometers cooled to 4.2 K at the synchrotron SOLEIL [183]. At ANKA a similar setup is intended using our fast YBCO THz detectors. For a successful

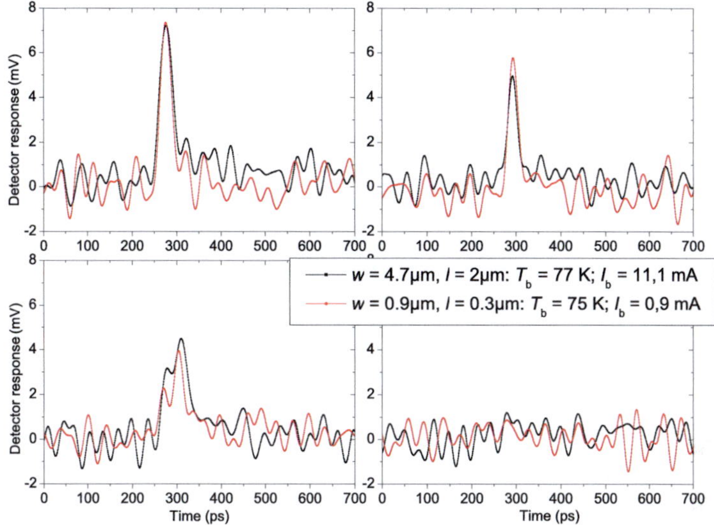

Fig. 7.15.: Single shots recorded with a microbridge and the smallest developed sub-µm detector excited by the same CSR THz pulse [Raa12]. The recorded transients show excellent agreement in shape and magnitude.

implementation of balanced detection, the used detectors must show identical response transients to the same excitation. To demonstrate that, two different YBCO detectors developed in this work were tested and compared at ANKA: a typical microbridge detector with $w = 4.7$ µm and $l = 2$ µm and the smallest developed sub-µm detector with $w = 0.9$ µm and $l = 0.3$ µm. The microbridge was installed in the liquid nitrogen cryostat while the sub-µm detector was operated in the cryogen-free system. The CSR THz radiation was split using a wire grid and the bias current of the sub-µm detector was adjusted to have the same amplitude compared to the microbridge. The different operation temperatures as well as the different biasing conditions are shown in Fig. 7.15. For these very different detector operation conditions multiple single shots of both detector responses were recorded shown in Fig. 7.15. The varying amplitudes due to the CSR bursting were recorded showing excellent agreement between the two detector pulse shapes as well as the detector amplitudes.

The successful demonstration of identical response transients to the same CSR THz pulse allows to realize a balanced detection system at the ANKA storage ring.

MLS - Metrology light source

The Metrology Light Source (MLS) [170] has a circumference of 48 m and can be operated at any electron beam energy between 105 and 630 MeV. Moreover, it is optimized for the generation of coherent synchrotron radiation in the far IR/ THz range. The stored electron beam current can be varied by more than 11 decades from a maximum current of 200 mA down to one stored electron (1 pA).

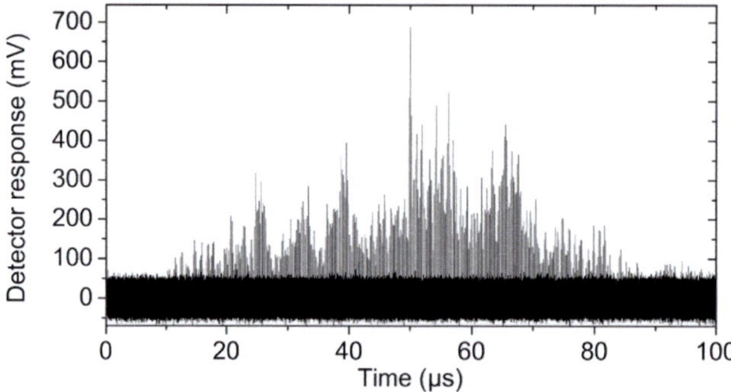

Fig. 7.16.: Detector response to a CSR single bunch filling over multiple turns obtained at UVSOR-II [TSW+13]. The strong influence of the wakefield of the electrons on the amplitude of the detector response can be seen.

A characteristic feature of all storage rings is the jitter in the arrival time of the electron bunch as a consequence of a slightly unstable orbit or longitudinal oscillations. The exact value of the jitter is determined by the storage rings settings and was studied at MLS with our YBCO thin-film detectors. For this the transients of the THz pulses were recorded and the time variation at half the rising edge of the pulses was extracted. It was found that the jitter shows a Gaussian distribution and is in the range of a few picoseconds. The jitter increases with increasing bunch current and also with increasing bunch length. A detailed analysis can be found in [PSH+12a].

UVSOR-II - Ultraviolet Synchrotron Radiation Facility

UVSOR-II, the electron storage ring of the Institute for Molecular Science in Okazaki, Japan, is operated in the beam energy range between 600 and 750 MeV. The ring has a circumference of 53.2 m. The radio frequency and the revolution frequency of the ring are 90.1 and 5.6 MHz,

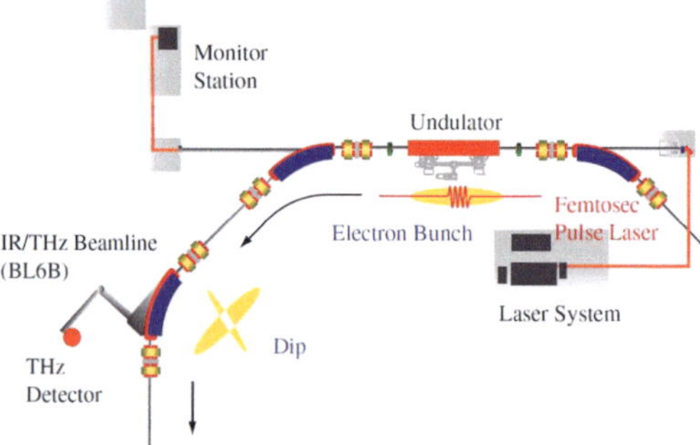

Fig. 7.17.: Schematic drawing of the experimental setup at UVSOR-II for generating laser-induced CSR [175]. The femtosecond laser pulse (red) interacts with the electron bunch (yellow) within the undulator generating CSR which is detected and analyzed at the THz beamline (BL6B).

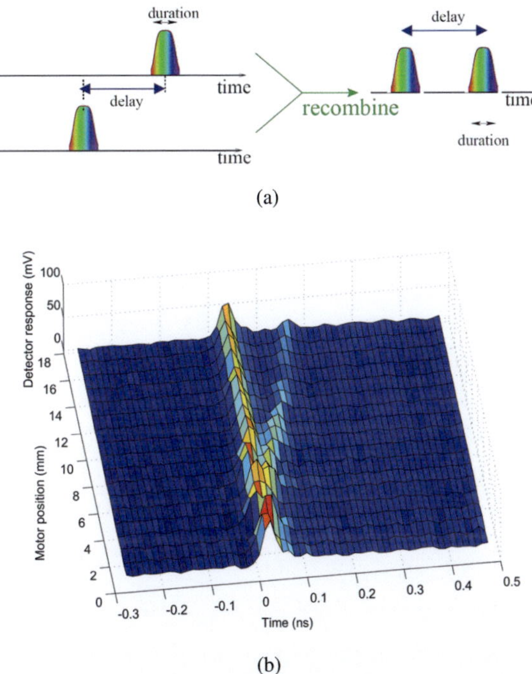

(a)

(b)

Fig. 7.18.: (a) Schematic of the laser pulse configuration: The delay between the 6 ps long pulses is adjusted by the Michelson mirror position allowing to generate two separated pulses in the THz frequency range [186]. (b) Detector response to THz CSR generated by the laser-slicing technique. The laser was guided through a Michelson interferometer, thus leading to a continuous separation of the detector pulses with increasing mirror position.

respectively. UVSOR-II has the possibility to work in top-up operation which allows to continuously refill electrons in the beam, thus allowing to keep the electron beam current constant.

In Fig. 7.16 the YBCO detector response to a single bunch filling recorded over 100 µs is displayed. The beam was operated at 600 MeV above the beam current bursting threshold of 50 mA, so bursting coherent synchrotron radiation was emitted. The interaction of the electrons with their own field, i.e. the influence of the wakefield, results in a strong amplitude variation of the emitted pulses (see Fig. 7.16). For the analysis of these bursting dynamics ultra-fast detectors are required which are able to follow the single picosecond pulses emitted by the storage ring.

Another feature of the UVSOR-II storage ring, is the possibility to generate laser-induced coherent THz radiation as discussed above. Relativistic electrons show striking non-linear collective behaviours when interacting with an intense laser beam, which can lead to powerful laser-induced coherent emission [184, 185]. The setup at UVSOR-II to generate laser-induced CSR is displayed in Fig. 7.17.

For the measurements shown in Fig. 7.18(b) the laser-slicing technique has been employed. A 6 ps long laser pulse was guided through a Michelson interferometer, adjusting the delay by the

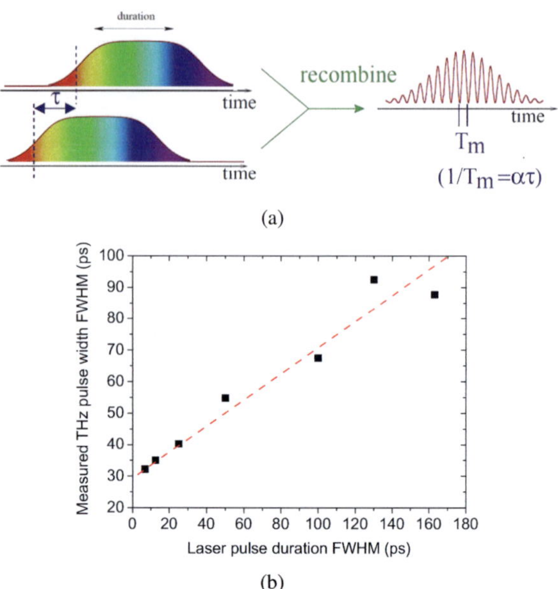

(a)

(b)

Fig. 7.19.: (a) The pulse duration of the laser pulse is much larger than its separation. The output is a pulse with a sine modulation with a modulation frequency proportional to the delay [186]. (b) Dependence of measured THz pulse width in dependence on the laser pulse duration. The measured THz pulse width decreases linearly with decreasing laser pulse duration.

motor position of the mirror over 18 mm (see Fig. 7.18(a)), before feeding the laser pulses into the storage ring. The continuous increase in separation between the two detector transients measured in the THz frequency range, generated by the optical laser pulses, is clearly visible. This was the first time that this phenomena could be observed at UVSOR-II in Japan since a detector with picosecond time resolution is required for this experiment.

As already discussed a characteristic feature of CSR emission in electron storage rings is the broadband emission of radiation from 0.2 - 2 THz, which is for some measurements disadvantageous. In 2008, Bielawski *et al.* [177] demonstrated the possibility of mastering the coherent emission experimentally by producing narrowband terahertz radiation via two laser pulses which are overlayed resulting in modulated laser pulse. In Fig. 7.19(a) the general idea of the laser pulse modulation is displayed. This laser pulse, now possessing a longitudinal quasi-sinusoidal modulation, interacts with the electron bunch of the storage ring in the undulator (see Fig. 7.17) leading to narrowband emission of THz CSR. The emission frequency is equal to the laser-pulse modulation frequency and the pulse duration can be varied from a few picoseconds to 100 ps [177]. Fig. 7.19(b) shows measurements with our YBCO detector of this narrowband generated CSR at a modulation frequency of 150 GHz. A linear decrease of the measured THz pulse width with decreasing laser pulse duration was observed proofing the operation principle.

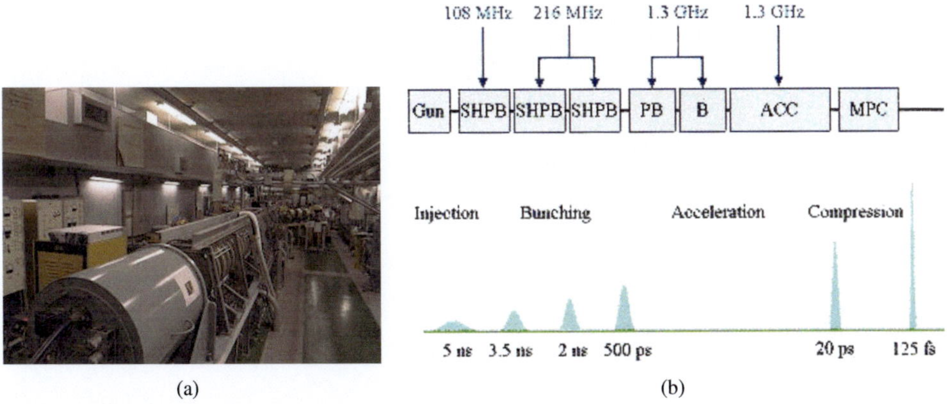

Fig. 7.20.: (a) Photograph of the L-band electron linac at Osaka University. (b) Schematic of the generation of coherent high-power THz pulsed radiation in the L-band electron linac.

7.2.2. Free-electron laser - The L-band linear accelerator

A free-electron laser, or FEL, is a laser that shares the same optical properties as conventional lasers such as emitting a beam consisting of coherent electromagnetic radiation which can reach high power [187]. However, the FEL uses some very different operating principles to form the beam. Unlike gas, liquid, or solid-state lasers such as diode lasers, in which electrons are excited in bound atomic or molecular states, FELs use a relativistic electron beam as the lasing medium which moves freely through a magnetic structure, hence the term free electron [187]. The free-electron laser has the widest frequency range of any laser type, and can be widely tunable [188], currently ranging in wavelength from microwaves, through terahertz radiation, infrared and the visible spectrum, to ultraviolet and X-rays. To create a FEL, a beam of electrons is accelerated to almost the speed of light. The beam passes through the FEL oscillator, a periodic transverse magnetic field produced by an arrangement of magnets with alternating poles within an optical cavity along the beam path. This array of magnets is called an undulator, or a "wiggler", because it forces the electrons in the beam to follow a sinusoidal path. The acceleration of the electrons along this path results in the release of photons (synchrotron radiation). Since the electron motion is in phase with the field of the light already emitted, the fields add together coherently [189]. The wavelength of the light emitted can be readily tuned by adjusting the energy of the electron beam or the magnetic field strength of the undulators.

The L-band electron linear accelerator (linac) (see Fig. 7.20) at the Institute of Scientific and Industrial Research at the Department of Accelerator Science at the Osaka University [190] was constructed in 1978 for generating intense electron beam singly bunched in picosecond width. The charge of electrons in the single bunch beam amounts to 91 nC. The high-brightness electron beam has been mainly used for studies of the transient phenomena in the range from nanoseconds to sub-picoseconds by means of pulse radiolysis. Further applications are the development of a far-infrared FEL as well as basic research on Self-Amplified Spontaneous Emission (SASE) in the

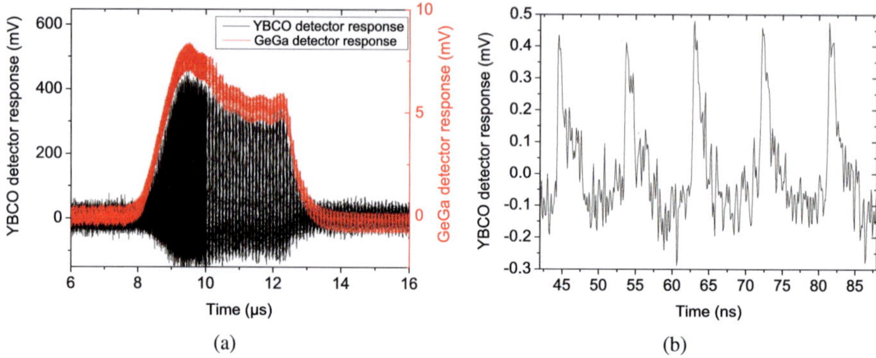

Fig. 7.21.: (a) YBCO detector response to the emitted THz radiation of the linac compared to the detector response of a Ge:Ga photoconductive switch. The Ge:Ga detector can only measure the envelope of the filling pattern, whereas the YBCO detector follows every peak. (b) Zoom of the YBCO detector response to single pulses with a repetition time of 9.2 ns.

far-infrared region [190]. The linac was renewed in 2003 for the purpose of realizing high stability and reproducibility as well as easy operation. The linac is composed of a thermionic electron gun, three subharmonic prebunchers (SHPBs), a pre-buncher (PB), a buncher (B) and a 3 m long accelerating tube. Electrons injected from the gun and passing through the SHPB system are bunched to 20 - 30 ps by PB and B and accelerated up to 40 MeV through the acceleration tube driven by a klystron with the peak power of 30 MW at the rf-frequency of 1.3 GHz.

In this work, we report measurements of coherent THz radiation produced by the L-band linac in collaboration with the group of Prof. Isoyama from the Institute of Scientific and Industrial Research at the Department of Accelerator Science. Their research focus is on the production of highly brilliant electron beams with the linac and related beam dynamics as well as the development of a far-infrared or THz FEL for user experiments. For the detailed study and analysis of the generated radiation a very fast detector with a very broad dynamic range is required. Thus, the collaboration was established and measurements with our YBCO detectors were carried out at the L-band linac in Japan which are discussed below.

In Fig. 7.21 the YBCO detector response to the coherent THz radiation produced by the L-band linac is shown. For the characterization of the emitted radiation a Ge:Ga photoconductive switch is used at the group of Prof. Isoyama. In Fig. 7.21(a) the Ge:Ga detector response is compared to the response of our YBCO thin-film detector. The Ge:Ga detector is only able to measure the envelope of the emitted radiation pattern, while the YBCO detector is able to resolve the single peaks (see Fig. 7.21(b)) thus allowing to study inter-bunch influences. The low temporal resolution of the single pulses is explained by the real-time oscilloscope with a bandwidth of 1 GHz which was available.

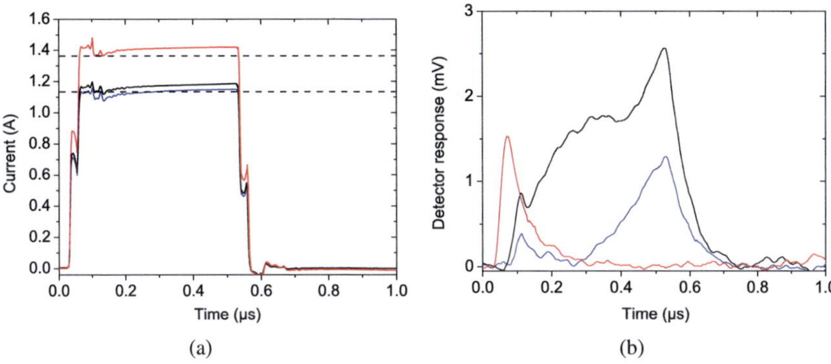

Fig. 7.22.: (a) Measured QCL current as a function of time for different bias voltages: $U_b = 12.2$ V (blue), $U_b = 12.6$ V (black), $U_b = 14.8$ V (red). The dashed horizontal lines indicate the threshold and cut-off current of the laser. (b) YBCO detector response for the corresponding QCL bias voltages.

7.2.3. Quantum cascade laser

Terahertz quantum cascade lasers (THz-QCLs) are powerful, compact, solid-state sources of coherent THz radiation over a frequency range from 1.2 to 5 THz [191, 192]. The high output power (exceeding 250 mW in pulsed operation [193]), narrow linewidths (<30 kHz [194]) and broad range of achievable emission frequencies give rise to numerous potential applications [195]. To optimize these THz sources a clear understanding of the dynamic processes inherent in these devices are required, which can occur on fast time-scales owing to the lifetime of the upper state of the lasing transition being limited to a few picoseconds by elastic and inelastic scattering mechanisms in THz-QCLs [196].

During this work a collaboration with the School of Electronic and Electrical Engineering at the University of Leeds was established to employ our developed YBCO detectors for the analysis of the dynamic behaviour of the emitted QCL radiation which was not possible with the available Ge bolometer due to its millisecond response time. A detailed discussion of the obtained results is given in [SDV[+]13].

Fig. 7.22(a) shows the time-dependent current flow for different values of the QCL bias voltage U_b, for a pulse length of 500 ns. A QCL emits radiation when its drive current exceeds the threshold level and is below the cut-off level determined by the waveguide and mirror losses, the overlap of the waveguide mode with the active region, and the laser gain coefficient. For the analyzed laser, the threshold current ($I_{th} = 1.13$ A) and cut-off current ($I_{cut-off} = 1.36$ A) are indicated by the dashed horizontal lines in Fig. 7.22(a). As the value of the QCL bias voltage U_b is increased as given in the caption of Fig. 7.22, the different curves show an increase in the maximum QCL current but maintain the same general pulse shape. In contrast to the QCL current, the measured detector signals show a strong variation in both the amplitude and shape of the pulses, as shown in Fig. 7.22(b). For higher QCL bias voltages U_b the threshold is earlier reached than for lower biases, and therefore the QCL starts emitting radiation earlier after the start of the current pulse.

This can be seen by comparing the $U_b = 14.8$ V and the $U_b = 12.6$ V traces in Fig. 7.22. The first peak of the detector signal appears ≈ 40 ns earlier for the 14.8 V trace than for the 12.6 V trace (Fig. 7.22(b)) because the corresponding current traces reach the threshold level at different points in time (Fig. 7.22(a)) as a consequence of the specific transient features in these current pulses. Using our fast YBCO detectors it is possible to resolve such features thus demonstrating the ability to contribute to the understanding of the QCL dynamic processes as discussed in detail in [SDV$^+$13].

7.3. Conclusions

The measurements discussed in this chapter demonstrate the successful development of a YBCO thin-film detection system for time-domain measurements on a picosecond time scale in the THz frequency range.

The developed detectors were embedded in an ultra-fast readout system as discussed in section 7.1. The cryogen-free cryocooler-based system allows for permanent installation and upgrading to a fully automated detection system.

The maximum achieved effective readout bandwidth was 22 GHz corresponding to a minimum temporal resolution of only 16 ps (FWHM) which already covers a wide range of emitted THz pulses e.g. by an electron storage ring. For the first time, the real-time temporal evolution of emitted THz CSR pulses at the electron storage ring ANKA was successfully resolved. Full widths at half maximum of 17 ps were achieved which were in perfect agreement with the calculated width of the emitted CSR pulse.

With this developed ultra-fast YBCO detection system first studies for the analysis of the dynamics of relativistic electrons in electron storage rings and free-electron lasers were carried out (section 7.2). Furthermore, the dynamics of QCL transients were detected and analyzed with our detectors. All these experiments emphasize the recently emerged requirement for temporal resolutions on a picosecond time scale of direct THz detector technologies which was met by the development of the YBCO thin-film detection system in this work.

However, to resolve even the shortest THz pulses on the single picosecond time scale possible to generate e.g. by an electron storage ring, two strategies should be taken into account: either the effective readout bandwidth needs to be increased or optical readout techniques like an autocorrelation setup need to be employed. The bottleneck of the first approach is defined by the real-time oscilloscope in the readout. Beside the fact, that there is currently no real-time oscilloscope available with a bandwidth above 63 GHz, this component is also very expensive. Much more convenient, thus, is the implementation of an autocorrelation setup with which time resolutions below 1 ps are achievable. However, the author would like to mention that for an autocorrelation measurement a non-linear device is required. If this holds true for our YBCO thin-film detectors in the THz frequency range requires further investigations.

8. Summary

In this thesis, we have developed a new direct THz detector technology based on the high-temperature superconductor YBCO which opens new routes in the analysis of ultra-fast, picosecond time-domain processes over a broad THz frequency range (0.07 - 2 THz) with a wide dynamic range of more than 30 dB. This was successfully demonstrated by the recording of a 17 ps wide (FWHM) coherent synchrotron radiation (CSR) THz pulse at the electron storage ring ANKA. Furthermore, for the first time clear evidences for a vortex-assisted detection mechanism in YBCO thin-film detectors for pulsed photon excitations below the superconducting energy gap were found.

The motivation for the development of the new direct THz detector technology was driven by the recent enhancement of brilliant pulsed THz sources. Accelerator-based sources are able to provide brilliant THz pulses on a picosecond time scale. For the optimization of the emission of this high-power pulsed THz radiation, direct detectors are required which can resolve picosecond temporal processes in the THz frequency range. We chose the high-temperature superconductor YBCO as candidate for this detector application since in the optical frequency range very fast relaxation processes (single picoseconds) were already reported.

For the detector development we have optimized the deposition process for high-T_c YBCO thin-films with the focus on thicknesses below 30 nm required for detector applications. By the introduction and optimization of CeO_2 and PBCO buffer and protection layers, high critical temperatures above the boiling temperature of liquid nitrogen even for ultra-thin films of only 10 nm thickness, corresponding to 8 unit cells, have been achieved. Critical temperatures of 79 K allow us to cool even these ultra-thin detectors by liquid nitrogen.

A patterning technology for micrometer- and sub-micrometer-sized YBCO bridges embedded in a gold antenna has been developed. By electron-beam lithography and a double-step etching process (physical ion milling and chemical wet-etching) we have realized YBCO bridges with lengths as short as 300 nm. The DC characterization as well as the long-term stability study of the fabricated detectors revealed high-quality superconducting devices ($T_c \geq 83$ K, $j_c \geq 1.5$ MA/cm^2 at 77 K) suitable for a permanent application.

For a clear understanding of the detection mechanism in our developed YBCO detectors we have performed a detailed photoresponse study to over-gap and sub-gap excitations by optical and THz pulsed and continuous radiation.

In the optical frequency range, we have found good agreement of the studied photoresponse to over-gap excitations with the two-temperature model. The experimentally determined time constants for the electron-phonon interaction time as well as the phonon escape time were consistent with already published values. The ultra-fast electron phonon interaction time of only 3.9 ps confirmed the motivation to use the high-temperature superconductor YBCO for ultra-fast detector applications.

The photoresponse study was then extended to THz wavelengths, i.e. to excitations below the superconducting energy gap of YBCO, using continuous as well as pulsed radiation. The measurements with continuous THz radiation have been employed for the determination of the *NEP* and responsivity values of our detectors. For the sub-micrometer devices we have achieved *NEP* values as small as 12 pWHz$^{-0.5}$ at 10 kHz and responsivity values of 1600 V/W. The fact that the responsivity dependence on bias current was well reproduced by the standard bolometer model suggests that the YBCO bridges respond bolometrically to continuous THz radiation which can be explained by the strong heating of the bridge inherent to this measurement technique.

However, in the THz pulsed-excitation regime a clear deviation from the bolometric behaviour was observed. For this case of sub-gap excitations, it was not possible to explain the detector responses in the framework of the two-temperature model. The response amplitudes of the detector were an order of magnitude larger in the THz frequency range compared to the optical frequency range for the same absorbed energy level. However, despite the large amplitudes at THz wavelengths, we have not observed a nanosecond long bolometric component in the detector response which is characteristic for the two-temperature model. Furthermore, the dependence of the detector response on bias current and absorbed energy showed a linear dependence in contrast to the non-linear behaviour observed in the optical frequency range.

These experimental results evidence a different detection mechanism at THz wavelengths not caused by electron heating and subsequent electron and phonon cooling. This idea was supported by the fact of zero-bias detection at THz wavelengths. Whereas for the bolometric detection at optical wavelengths no zero-bias transients were observed, we have found a clear zero-bias response for pulsed THz excitations.

To prove the idea of a different detection mechanism we have studied the dependence of the YBCO detector response on the phase of the THz pulse electric field. The polarity of the phase of the incident THz electric field was switched by changing the metallic mirror, guiding the THz pulse to the detector, to a dielectric one which directly led to a switching of the polarity of the detector response. This clearly indicated that the incident electric field of the THz pulse causes the detector response instead of its intensity which would be the case for a bolometric response mechanism.

A possible model explaining the photoresponse at THz pulsed excitations is based on a vortex-assisted detection mechanism. The idea is that the absorbed THz radiation in the planar antenna is transformed into a rf-current. This current pulse generates a Lorentz force which causes a rearrangement of the vortex lattice into a vortex line, which creates a channel with reduced superconducting order parameter. Vortices cross the bridge via this channel giving, due to the dissipative movement of the vortices, the voltage transient. Such a scenario explains qualitatively the experimental data of pulsed sub-gap excitations discussed in this work.

For the readout of the fast picosecond detector pulses a new, ultra-fast detection system has been developed at IMS. The new YBCO detectors have been embedded in the detection system achieving a temporal resolution of 16 ps (FWHM). Successful measurements at several pulsed THz sources have been carried out. At ANKA, the synchrotron from KIT, the real-time evolution of the emitted THz pulses has been recorded for the first time resulting in pulse widths of 17 ps (FWHM). At the Ultraviolet Synchrotron Radiation Facility (UVSOR) in Japan we have recorded the bursting behaviour of CSR for high electron beam currents as well as laser-induced CSR in detail. Furthermore, the detailed filling pattern of the free-electron laser of the Osaka University has been resolved. Finally, we have studied the dynamic behaviour of quantum cascade lasers with our developed detectors.

To further optimize the YBCO THz detector technology a detailed model for the detection mechanism is required. Experiments in magnetic field generating vortices in the detector would lead to a controllable vortex state. A detailed study of the detector response to pulsed THz radiation in these controllable and variable vortex states would lead to new insights in the detection mechanism for sub-gap pulsed excitations. Furthermore, the photoresponse experiments should be extended to the sub-μm detector structures since the lateral dimensions will become of the order of the travel distance of the vortices within the superconducting bridge. This limiting case should influence on the detector amplitude as well as on the detector time constant giving further insight in the dynamic processes occurring in the YBCO detector for sub-gap pulsed excitations.
Finally, a detailed analysis of coherent terahertz radiation is now possible as well as the general implementation of the ultra-fast YBCO time-domain measurement setup for the detection of picosecond THz dynamic processes opening new routes in THz direct detector applications.

A. Electrical responsivity according to Jones

In this chapter, the mathematical conversion from the general responsivity equation of a bolometer (see equation A.1) to the responsivity equation according to Jones [161] is given. The electrical responsivity according to Jones is limited to the case of excitations with low modulation frequencies ω which is true for continuous THz radiation measurements described in section 6.2.1. For this case the equation gives a clear understanding of parameters influencing on the bolometer's responsivity which are the bolometer resistance R, the differential resistance $Z = dV/dI$ and the bias current I of the bolometer. In the following the conversion is described in detail.

As demonstrated in equation 2.8 the responsivity S for a bolometer is generally described by

$$S = \frac{\eta \alpha R I}{G_{eff}(1 + \omega^2 \tau^2)^{1/2}} \tag{A.1}$$

where η describes the absorptivity of the bolometer, α is the temperature coefficient of resistance, G_{eff} is the effective thermal conductance and τ the time constant of the bolometer.

For excitations with low modulation frequencies ω, $\omega\tau$ becomes $\ll 1$ for YBCO bolometers with a time constant τ in the nanosecond range. Together with $\eta = 1$ this simplifies equation A.1 to

$$S = \frac{\alpha R I}{G_{eff}} = \frac{\alpha V}{G - \alpha P} \tag{A.2}$$

with the bolometer voltage V, the thermal conductance G and the electrical power P.
Eliminating dT from the following expressions

$$dV = d(IR) = RdI + V\alpha dT \tag{A.3}$$

$$dP = GdT = d(IV) = VdI + IdV \tag{A.4}$$

results in

$$dV = RdI + V\alpha\frac{VdI + IdV}{G} = RdI + \frac{V^2\alpha dI + \alpha P dV}{G}. \tag{A.5}$$

This is further transformed to

$$dVG = RGdI + V^2\alpha dI + \alpha P dV \tag{A.6}$$

and

$$dV(G - \alpha P) = dI(RG + V^2\alpha). \tag{A.7}$$

This allows to express the differential resistance Z as

$$Z = \frac{dV}{dI} = \frac{RG - V^2\alpha}{G - \alpha P} = \frac{R(G + \alpha P)}{G - \alpha P} \tag{A.8}$$

and the thermal conductance in dependence of the differential resistance

$$Z(G - \alpha P) = R(G + \alpha P) \tag{A.9}$$

$$ZG - \alpha PZ = RG + R\alpha P \tag{A.10}$$

$$G(Z - R) = R\alpha P + Z\alpha P \tag{A.11}$$

$$G = \frac{\alpha P(R + Z)}{Z - R}. \tag{A.12}$$

By replacing now G in equation A.2 an equation of the bolometer responsivity only dependent on values derivable from the superconducting transition and IV characteristic is obtained:

$$S = \frac{\alpha V}{\frac{\alpha P(R + Z)}{(Z - R)} - \alpha P} \tag{A.13}$$

$$S = \frac{\alpha V(Z - R)}{\alpha P(R + Z) - \alpha P(Z - R)} \tag{A.14}$$

$$S = \frac{\alpha V(Z - R)}{\alpha PR + \alpha PZ - \alpha PZ + \alpha PR} \tag{A.15}$$

$$S = \frac{\alpha V(Z - R)}{2\alpha PR} = \frac{V(Z - R)}{2IVR} \tag{A.16}$$

$$S = \frac{Z - R}{2IR} \tag{A.17}$$

B. Energy relaxation of superconducting thin films according to Perrin and Vanneste

In this chapter, the response of a superconducting thin film to periodic optical irradiation $P_{in}(t)$ according to Perrin and Vanneste [105] is discussed. The general solution of the 2T model is simplified for the case of excitations of YBCO thin films showing that the film voltage response can be expressed by a single roll-off function.

The 2T model with the electron temperature T_e and phonon temperature T_p is described by the linearized differential equations as discussed in section 3.2.1 which are

$$\frac{dT_e}{dt} = -\frac{T_e - T_p}{\tau_{ep}} + \frac{1}{C_e}\frac{\alpha P_{in}(t)}{V} \tag{B.1}$$

$$\frac{dT_p}{dt} = -\frac{C_e}{C_p}\frac{T_e - T_p}{\tau_{ep}} - \frac{T_p - T_b}{\tau_{es}} \tag{B.2}$$

with $W(t) = \alpha P_{in}(t)/V$ where α is the radiation absorption coefficient and V the volume of the superconducting thin film. The electron heat capacity C_e and phonon heat capacity C_p together with the electron-phonon interaction time τ_{ep} and the phonon escape time τ_{es} characterize the electron and phonon subsystem of the superconducting thin film.

The change of the electron temperature ΔT_e excited by the absorbed radiation power can be measured by the change of voltage ΔU due to a change of the temperature dependent resistance $R(T)$ of the sample according to

$$\Delta U = I_{bias}\frac{dR}{dT}\Delta T_e. \tag{B.3}$$

For a periodic excitation of $P_{in}(t) = P_0\cos\omega t$ the change of the electron temperature ΔT_e was derived from the differential equations of the 2T model (see equation B.1 and B.2) by Perrin and Vanneste [105] and results in the voltage change

$$\Delta U = I_{bias}\frac{dR}{dT}\frac{\alpha P_0\tau_{es}}{(C_e+C_p)V}\sqrt{\frac{1+[\omega\tau_{ep}(1+C_p/C_e)]^2}{[1+(\omega\tau_{es})^2][1+(\omega\tau_{ep})^2]}}. \tag{B.4}$$

As discussed in section 3.2.1, for the high-temperature superconductor YBCO the electron-phonon interaction time is much smaller than the phonon escape time. With this assumption $\tau_{ep} \ll \tau_{es}$ and the assumption of low frequencies ω, equation B.4 can be simplified to

$$\Delta U(\Delta\omega) = I_{bias}\frac{dR}{dT}\frac{\alpha P_0\tau_{es}}{(C_e+C_p)V}\sqrt{\frac{1}{[1+(\Delta\omega\tau_{es})^2]}} = \frac{\Delta U(0)}{\sqrt{1+(\Delta\omega/\omega_{es})^2}}. \tag{B.5}$$

This single roll-off function was used in section 6.1 to fit the photoresponse data of the YBCO thin film detectors with the fitting parameter $\Delta U(0)$ and $f_{es} = \omega_{es}/2\pi$.

List of Tables

Bibliography

[1] A.-S. Müller. Accelerator-Based Sources of Infrared and Terahertz Radiation. *Reviews of Accelerator Science and Technology*, 3(1):165–183, 2010.

[2] P. H. Siegel. Terahertz Technology. *IEEE Transactions on Microwave theory and techniques*, 50(3):910–936, 2002.

[3] W. D. Duncan and G. P. Williams. Infrared synchrotron radiation from electron storage rings. *Applied Optics*, 22(18):2914–2923, 1983.

[4] M. Abo-Bakr, J. Feikes, K. Holldack, G. Wüstefeld, and H.-W. Hübers. Steady-State Far-Infrared Coherent Synchrotron Radiation detected at BESSY II. *Physical Review Letters*, 88(25):254801, 2002.

[5] A.-S. Müller, Y.-L. Mathis, I. Birkel, B. Gasharova, C. J. Hirschmugl, E. Huttel, D. A. Moss, R. Rossmanith, and P. Wesolowski. Far Infrared Coherent Synchrotron Edge Radiation at ANKA. *Synchrotron Radiation News*, 19(3):18–24, 2006.

[6] F. Wang, D. Cheever, M. Farkhondeh, W. Franklin, E. Ihloff, J. van der Laan, B. McAllister, R. Milner, C. Tschalaer, D. Wang, D. F. Wang, A. Zolfaghari, T. Zwart, G. L. Carr, B. Podobedov, and F. Sannibale. Coherent THz Synchrotron Radiation from a Storage Ring with High-Frequency RF System. *Physical Review Letters*, 96(6):064801, 2006.

[7] R. Müller, A. Hoehl, A. Matschulat, A. Serdyukov, G. Ulm, J. Feikes, M. Ries, and G. Wüstefeld. Status of the IR and THz beamlines at the Metrology Light Source. *Journal of Physics: Conference Series*, 359(012004):1–8, 2012.

[8] A. Plech, S. Casalbuoni, B. Gasharova, E. Huttel, Y.-L. Mathis, A.-S. Müller, K. Sonnad, A. Bartels, and R. Weigel. Electro-optical sampling of terahertz radiation emitted by short bunches in the ANKA synchrotron. *Proceedings of PAC09, Vancouver, BC, Canada*, TU5RFP026:1–3, 2009.

[9] A. D. Semenov, R. S. Nebosis, Y. P. Gousev, M. A. Heusinger, and K. F. Renk. Analysis of the nonequilibrium photoresponse of superconducting films to pulsed radiation by use of a two-temperature model. *Physical Review B*, 52(1):581–590, 1995.

[10] M. Danerud, D. Winkler, M. Lindgren, M. Zorin, V. Trifonov, B. S. Karasik, G. N. Gol'tsman, and E. M. Gershenzon. Nonequilibrium and bolometric photoresponse in patterned $YBa_2Cu_3O_{7-\delta}$ thin films. *Journal of Applied Physics*, 76(3):1902–1909, 1994.

[11] M. Lindgren, M. Currie, C. A. Williams, T. Y. Hsiang, P. M. Fauchet, R. Sobolewski, S. H. Moffat, R. A. Hughes, J. S. Preston, and F. A. Hegmann. Ultrafast photoresponse in microbridges and pulse propagation in transmission lines made from high-T_c superconducting Y-Ba-Cu-O thin films. *IEEE Journal of selected topics in quantum electronics*, 2(3):668–678, 1996.

[12] A. D. Semenov, I. G. Goghidze, G. N. Gol'tsman, A. V. Sergeev, and E. M. Gershenzon. Evidence for the spectral dependence of nonequilibrium picosecond photoresponse of YBaCuO thin films. *Applied Physics Letters*, 63(5):681–683, 1993.

[13] A. Rostami, H. Rasooli, and H. Baghban. *Terahertz Technology*. Springer Verlag Berlin-Heidelberg, 2011.

[14] H. Kraus. Superconductive bolometers and calorimeters. *Superconductor Science and Technology*, 9:827–842, 1996.

[15] A. T. Lee, P. L. Richards, S. W. Nam, B. Cabrera, and K. D. Irwin. A superconducting bolometer with strong electrothermal feedback. *Applied Physics Letters*, 69(12):1801–1803, 1996.

[16] S. M. Wentworth and D. P. Neikirk. Composite Microbolometers with Tellurium Detector Elements. *IEEE Transactions on microwave theory and techniques*, 40(2):196–201, 1992.

[17] N. S. Nishioka, P. L. Richards, and D. P. Woody. Composite bolometers for submillimeter wavelengths. *Applied Optics*, 17(10):1562–1567, 1978.

[18] P. L. Richards. Bolometers for infrared and millimeter waves. *Journal of Applied Physics*, 76(1):1–24, 1994.

[19] D. C. Alsop, C. Inman, A. E. Lange, and T. Wilbanks. Design and construction of high-sensitivity, infrared bolometers for operation at 300 mK. *Applied Optics*, 31(31):6610–6615, 1992.

[20] W. A. Holmes, J. J. Bock, B. P. Crill, T. C. Koch, W. C. Jones, A. E. Lange, and C. G. Paine. Initial test results on bolometers for the Planck high frequency instrument. *Applied Optics*, 47(32):5996–6008, 2008.

[21] Thomas Keating Ltd. QMC Instruments Ltd. http://www.terahertz.co.uk/, 2011.

[22] D. H. Andrews. *American Philosophical Yearbook*. The American Philosophical Society, 1938.

[23] D. H. Andrews, W. F. Brucksch Jr., W. T. Ziegler, and E. R. Blanchard. Attenuated Superconductors I. For Measuring Infra-Red Radiation. *Review of Scientific Instruments*, 13:281–292, 1942.

[24] A. Goetz. The Possible Use of Superconductivity for Radiometric Purposes. *Physical Review*, 55:1270–1271, 1939.

[25] J. Clarke, P. L. Richards, and N.-H. Yeh. Composite superconducting transition edge bolometer. *Applied Physics Letters*, 30(12):664–666, 1977.

[26] S. Anders, T. May, V. Zakosarenko, K. Peiselt, E. Heinz, M. Starkloff, G. Zieger, E. Kreysa, G. Siringo, and H.-G. Meyer. Cryogenic bolometers for astronomical observations in the sub-mm range. *Microelectronic Engineering*, 88:2205–2207, 2011.

[27] M. Kenyon, P. K. Day, C. M. Bradford, J. J. Bock, and H. G. Leduc. Background-limited membrane-isolated TES bolometers for far-IR/submillimeter direct-detection spectroscopy. *Nuclear Instruments and Methods in Physics Research*, 559(2):456–458, 2006.

[28] P. W. Kruse. Physics and applications of high-T_c superconductors for infrared detectors. *Semiconductor Science and Technology*, 5:S229–S239, 1990.

[29] S. Verghese, P. L. Richards, K. Char, and S. A. Sachtjen. Fabrication of an infrared bolometer with a high T_c superconducting thermometer. *IEEE Transactions on magnetics*, 27(2):3077–3080, 1991.

[30] J. C. Brasunas and B. Lakew. High T_c superconductor bolometer with record performance. *Applied Physics Letters*, 64(6):777–778, 1994.

[31] M. J. M. E. de Neville, M. P. Bruijn, R. de Vries, J. J. Wijnbergen, P. A. J. de Korte, S. Sánchez, M. Elwenspoek, T. Heidenblut, B. Schwierzi, W. Michalke, and E. Steinbeiss. Low noise high-T_c superconducting bolometers on silicon nitride membranes for far-infrared detection. *Journal of Applied Physics*, 82(10):4719–4726, 1997.

[32] S. P. Langley. The Bolometer. *Nature*, 25(627):14–16, 1881.

[33] W. H. Block and O. L. Caddy. Thin-metal film room-temperature IR bolometers with nanosecond response times. *IEEE Journal of Quantum Electronics*, QE-9(11):1044–1053, 1973.

[34] G. M. Rebeiz, C. C. Ling, and D. B. Rutledge. Large-area bolometers for millimeter-wave power calibration. *International journal of Infrared and Millimeter Waves*, 10(8):931–928, 1989.

[35] V. S. Jagtap, A. Scheuring, M. Longhin, A. J. Kreisler, and A. F. Dégardin. From Superconducting to Semiconducting YBCO Thin Film Bolometers: Sensitivity and Crosstalk Investigations for Future THz Imagers. *IEEE Transactions on Applied Superconductivity*, 19(3):287–292, 2009.

[36] D. Murphy, M. Ray, R. Wyles, J. Asbrock, N. Lum, J. Wyles, C. Hewitt, A. Kennedy, and D. Van Lue. High sensitivity 25 μm microbolometer FPAs. *Proceedings of SPIE*, 4721:99–110, 2002.

[37] E. Mottin, A. Bain, J. L. Martin, J. L. Ouvrier-Buffet, J. J. Yon, J. P. Chatard, and J. L. Tissot. Uncooled amorphous silicon technology: high performance achievement and future trends. *Proceedings of SPIE*, 4721:56–63, 2002.

[38] H. Wada, T. Sone, H. Hata, Y. Nakaki, O. Kaneda, Y. Ohta, M. Ueno, and M. Kimata. YBaCuO uncooled microbolometer IRFPA. *Proceedings of SPIE*, 4369:299–304, 2001.

[39] F. J. Low. Low-temperature Germanium bolometer. *Journal of the optical society of America*, 51(11):1300–1304, 1961.

[40] U. P. Oppenheim, M. Katz, G. Koren, E. Polturak, and M. R. Fishman. High-temperature superconducting bolometer. *Physica C*, 178:26–28, 1991.

[41] G. L. Carr, M. Quijada, D. B. Tanner, C. J. Hirschmugl, G. P. Williams, S. Etemad, B. Dutta, F. DeRosa, A. Inam, T. Venkatesan, and X. Xi. Fast bolometric response by high T_c detectors measured with subnanosecond synchrotron radiation. *Applied Physics Letters*, 57(25):2725–2727, 1990.

[42] J. C. Brasunas, S. H. Moseley, B. Lakew, R. H. Ono, D. G. McDonald, J. A. Beall, and J. E. Sauvageau. Construction and performance of a high-temperature-superconductor composite bolometer. *Journal of Applied Physics*, 66(9):4551–4554, 1989.

[43] A. J. Kreisler and A. Gaugue. Recent progress in HTSC bolometric detectors at terahertz frequencies. *Proceedings of SPIE*, 3481:457–467, 1998.

[44] Z. M. Zhang and A. Frenkel. Thermal and Nonequilibrium Responses of Superconductors for Radiation Detectors. *Journal of Superconductivity*, 7(6):871–884, 1994.

[45] T.-L. Hwang, S. E. Schwarz, and D. B. Rutledge. Microbolometers for infrared detection. *Applied Physics Letters*, 34(11):773–776, 1979.

[46] D. P. Neikirk, W. W. Lam, and D. B. Rutledge. Far-infrared microbolometer detectors. *International Journal of Infrared and Millimeter Waves*, 5(3):245–278, 1984.

[47] S. E. Schwarz and B. T. Ulrich. Antenna-coupled infrared detectors. *Journal of Applied Physics*, 48(5):1870–1873, 1977.

[48] T. Shimizu, H. Moritsu, Y. Yasuoka, and K. Gamo. Fabrication of antenna-coupled microbolometers. *Japanese Journal of Applied Physics*, 34(12A):6352–6357, 1995.

[49] E. N. Grossman, J. E. Sauvageau, and D. G. McDonald. Lithographic spiral antennas at short wavelengths. *Applied Physics Letters*, 59(25):3225–3227, 1991.

[50] M. E. MacDonald and E. N. Grossman. Niobium Microbolometers for Far-Infrared Detection. *IEEE Transactions on Microwave Theory and Techniques*, 43(4):893–896, 1995.

[51] A. Rahman, E. Duerr, G. de Lange, and Q. Hu. Micromachined room-temperature microbolometer for mm-wave detection and focal-plane imaging arrays. *Proceedings of SPIE*, 3064:122–133, 1997.

[52] A. Luukanen, V.-P. Viitanen, G. de Lange, and Q. Hu. Terahertz imaging system based on antenna-coupled microbolometers. *Proceedings of SPIE*, 3378:34–44, 1998.

[53] S. Cherednichenko, A. Hammar, S. Bevilacqua, V. Drakinskiy, J. Stake, and A. Kalabukhov. A Room Temperature Bolometer for Terahertz Coherent and Incoherent Detection. *IEEE Transactions on Terahertz Science and Technology*, 1(2):395–402, 2011.

[54] J. Nemarich. Microbolometer Detectors for Passive Millimeter-Wave Imaging. *Army Research Laboratory*, TR-3460:1–33, 2005.

[55] M. Nahum and P. L. Richards. Design analysis of a novel low temperature bolometer. *IEEE Transactions on Magnetics*, 27(2):2484–2487, 1991.

[56] S. Wentworth and D. Neikirk. The transition-edge microbolometer. *Proceedings of SPIE*, 1292:148–154, 1990.

[57] A. Luukanen and J. P. Pekola. A superconducting antenna-coupled hot-spot microbolometer. *Applied Physics Letters*, 82(22):3970–3972, 2003.

[58] A. D. Semenov, H.-W. Hübers, K. S. Il'in, M. Siegel, V. Judin, and A.-S. Müller. Monitoring Coherent THz-Synchrotron Radiation with Superconducting NbN Hot-Electron Detector. *Proceedings of 34th International Conference on Infrared, Millimeter, and Terahertz Waves*, 2009.

[59] W. Zhang, W. Miao, K. M. Zhou, S. L. Li, Z. H. Lin, Q. J. Yao, and S. C. Shi. Heterodyne Mixing and Direct Detection Performance of a Superconducting NbN Hot-Electron Bolometer. *IEEE Transactions on applied superconductivity*, 21(3):624–627, 2011.

[60] D. F. Santavicca, M. O. Reese, A. B. True, C. A. Schmuttenmaer, and D. E. Prober. Antenna-Coupled Niobium Bolometers for Terahertz Spectroscopy. *IEEE Transactions on applied superconductivity*, 17(2):412–415, 2007.

[61] Q. Hu and P. L. Richards. Design analysis of a high T_c superconducting microbolometer. *Applied Physics Letters*, 55(23):2444–2446, 1989.

[62] M. Nahum, Q. Hu, P. L. Richards, S. A. Sachtjen, N. Newman, and B. F. Cole. Fabrication and measurement of high-T_c superconducting microbolometers. *IEEE Transactions on magnetics*, 27(2):3081–3084, 1991.

[63] J. P. Rice, E. N. Grossman, and D. A. Rudman. Antenna-coupled high-T_c air-bridge microbolometer on silicon. *Applied Physics Letters*, 65(6):773–775, 1994.

[64] C. Barholm-Hansen and M. T. Levinsen. A HTC Microbolometer for far infrared detection. *Proceedings of SPIE*, 1512:218–226, 1991.

[65] I. A. Khrebtov, V. N. Leonov, A. D. Tkachenko, P. V. Bratukhin, A. A. Ivanov, A. V. Kuznetsov, H. Neff, and E. Steinbeil. Noise of high-T_c superconducting bolometers. *Proceedings of SPIE*, 3287:288–299, 1998.

[66] A. Hammar, S. Cherednichenko, S. Bevilacqua, V. Drakinskiy, and J. Stake. Terahertz direct detection in $YBa_2Cu_3O_7$ Microbolometers. *IEEE Transactions on Terahertz Science and Technology*, 1(2):390–394, 2011.

[67] F. Sizov. THz radiation sensors. *Opto-Electronics Review*, 18(1):10–36, 2010.

[68] T. W. Crowe, W. L. Bishop, D. W. Porterfield, J. L. Hesler, and R. M. Weikle II. Opening the terahertz window with integrated diode circuits. *IEEE Journal of solid-state circuits*, 40(10):2104–2110, 2005.

[69] V. G. Bozhkov. Semiconductor detectors, mixers, and frequency multipliers for the terahertz band. *Radiophysics and Quantum Electronics*, 46(8):631–656, 2003.

[70] A. Rogalski and F. Sizov. Terahertz detectors and focal plane arrays. *Opto-Electronics Review*, 19(3):346–404, 2011.

[71] D. T. Young and J. C. Irvin. Millimeter frequency conversion using Au-n-type GaAs Schottky barrier epitaxial diodes with a novel contacting technique. *Proceedings of the IEEE*, 53(12):2130–2132, 1965.

[72] H. P. Röser, H.-W. Hübers, E. Bründermann, and M. F. Kimmitt. Observation of mesoscopic effects in Schottky diodes at 300 K when used as mixers at THz frequencies. *Semiconductor Science and Technology*, 11:1328–1332, 1996.

[73] A. Kreisler. Submicron-size GaAs Schottky diodes for applications in the 1 - 10 THz range: Theoretical and experimental studies. *Proceedings of the Society of Photo-Optical Instrumentation Engineers*, 666:51–63, 1986.

[74] R. U. Titz, H. P. Röser, G. W. Schwaab, H. J. Neilson, P. A. Wood, T. W. Crowe, W. C. B. Peatman, J. Prince, B. S. Deaver, H. Alius, and G. Dodel. Investigation of GaAs Schottky

Barrier diodes in the THz range. *International Journal of Infrared and Millimeter Waves*, 11(7):809–820, 1990.

[75] T. W. Crowe, D. P. Porterfield, J. L. Hesler, W. L. Bishop, D. S. Kurtz, and K. Hui. Terahertz sources and detectors. *SPIE Proceedings*, 5790:271–280, 2005.

[76] L. Liu, J. L. Hesler, H. Xu, A. W. Lichtenberger, and R. M. WeikleII. A Broadband Quasi-Optical Terahertz Detector Utilizing a Zero Bias Schottky Diode. *IEEE Microwave and wireless components letters*, 20(9):504–506, 2010.

[77] H. Ito, F. Nakajima, T. Ohno, T. Furuta, T. Nagatsuma, and T. Ishibashi. InP-Based Planar-Antenna-Integrated Schottky-Barrier Diode for Millimeter- and Sub-Millimeter-Wave Detection. *Japanese Journal of Applied Physics*, 47(8):6256–6261, 2008.

[78] A. Semenov, O. Cojocari, H.-W. Hübers, F. Song, A. Klushin, and A.-S. Müller. Application of Zero-Bias Quasi-Optical Schottky-Diode Detectors for Monitoring Short-Pulse and Weak Terahertz Radiation. *IEEE Electron Device Letters*, 31(7):674–676, 2010.

[79] Virginia Diodes, Inc., 2011. http://vadiodes.com.

[80] Microtech Instruments inc., 2012. http://www.mtinstruments.com/.

[81] S. Hargreaves and R. A. Lewis. Terahertz imaging: materials and methods. *Journal of Materials Science: Materials in Electronics*, 18:S299–S303, 2007.

[82] N. Karpowicz, H. Zhong, J. Xu, K. I. Lin, J. S. Hwang, and X. C. Zhang. Nondestructive sub-THz imaging. *Proceedings of SPIE*, 5727:132–142, 2005.

[83] Gentec-EO USA, Inc., 2012. http://www.gentec-eo.com.

[84] A. Dobroiu, M. Yamashita, J. Xu, Y. N. Ohshima, Y. Morita, C. Otani, and K. Kawase. Terahertz imaging system based on a backward-wave oscillator. *Applied Optics*, 43(30):5637–5646, 2004.

[85] E. E. Haller, M. R. Hueschen, and P. L. Richards. Ge:Ga photoconductors in low infrared backgrounds. *Applied Physics Letters*, 34(8):495–497, 1979.

[86] H.-W. Hübers, A. Semenov, K. Holldack, U. Schade, G. Wüstefeld, and G. Gol'tsman. Time domain analysis of coherent terahertz synchrotron radiation. *Applied Physics Letters*, 87(184103):184103-1–184103-3, 2005.

[87] A. D. Semenov, G. N. Gol'tsman, and R. Sobolewski. Hot-electron effect in superconductors and its applications for radiation sensors. *Superconductor Science and Technology*, 15:R1–R16, 2002.

[88] Y. P. Gousev, G. N. Gol'tsman, A. D. Semenov, E. M. Gershenzon, R. S. Nebosis, M. A. Heusinger, and K. F. Renk. Broadband ultrafast superconducting NbN detector for electromagnetic radiation. *Journal of Applied Physics*, 75(7):3695–3697, 1994.

[89] J. G. Bednorz and K. A. Müller. Possible High T_c Superconductivity in the Ba-LaCu-O System. *Zeitschrift für Physik B Condensed Matter*, 64:189–193, 1986.

[90] M. K. Wu, J. R. Ashburn, C. T. Torng, P. H. Hor, R. L. Meng, C. Gao, Z. J. Huang, Y. Q. Wang, and C. W. Chu. Superconductivity at 93 K in a new mixed-phase YBCO compound system at ambient pressure. *Physical Review Letters*, 58:908–910, 1987.

[91] D. A. Bonn, J. E. Greedan, C. V. Stager, T. Timusk, M. G. Doss, S. L. Herr, K. Kamarás, and D. B. Tanner. Far-Infrared Conductivity of the High-T_c Superconductor $YBa_2Cu_3O_7$. *Physical Review Letters*, 58(21):2249–2250, 1987.

[92] K. Yvon and M. Francois. Crystal structures of high-T_c oxides. *Zeitschrift für Physik B Condensed Matter*, 76(4):413–444, 1989.

[93] W. Buckel and R. Kleiner. *Supraleitung - Grundlagen und Anwendungen, 7. Auflage*. Wiley-VCH Verlag und Co.KGaA, Boschstraße 12, 69469 Weinheim, Germany, 2013.

[94] G. Blatter, M. V. Feigel'mann, V. B. Geshkenbein, A. I. Larkin, and V. M. Vinokur. Vortices in high-temperature superconductors. *Reviews of Modern Physics*, 66(4):1125–1388, 1994.

[95] Z. Schlesinger, R. T. Collins, D. L. Kaiser, and F. Holtzberg. Superconducting Energy Gap and Normal-State Reflectivity of Single Crystal Y-Ba-Cu-O. *Physical Review Letters*, 59(17):1958–1961, 1987.

[96] X.-l. Wang, T. Nanba, M. Ikezawa, Y. Isikawa, K. Mori, K. Kobayashi, K. Kasai, K. Sato, and T. Fukase. Optical properties of polycrystalline $YBa_2Cu_3O_{7-\delta}$. *Japanese Journal of Applied Physics*, 26(8):L1391–L1393, 1987.

[97] I. Bozovic, D. Kirillov, A. Kapitulnik, K. Char, M. R. Hahn, M. R. Beasley, T. H. Geballe, Y. H. Kim, and A. J. Heeger. Optical measurements on oriented thin $YBa_2Cu_3O_{7-\delta}$ films: lack of evidence for excitonic superconductivity. *Physical Review Letters*, 59(19):2219–2221, 1987.

[98] J. Heremans, D. T. Morelli, G. W. Smith, and S. C. Strite III. Thermal and electronic properties of rare-earth $Ba_2Cu_3O_x$ superconductors. *Physical review B*, 37(4):1604–1610, 1988.

[99] S. E. Inderhees, M. B. Salamon, T. A. Friedmann, and D. M. Ginsberg. Measurement of the specific-heat anomaly at the superconducting transition of $YBa_2Cu_3O_{7-\delta}$. *Physical review B*, 36(4):2401–2403, 1987.

[100] G. A. Thomas, H. K. Ng, A. J. Millis, R. N. Bhatt, R. J. Cava, E. A. Rietman, D. W. Johnson, Jr., G. P. Espinosa, and J. M. Vandenberg. Far-infrared spectra of polycrystalline $Ba_2YCu_3O_{9-d}$. *Physical Review B*, 36(1):846–849, 1987.

[101] J. R. Kirtley, R. T. Collins, Z. Schlesinger, W. J. Gallagher, R. L. Sandstrom, T. R. Dinger, and D. A. Chance. Tunneling and infrared measurements of the energy gap in the high-critical-temperature superconductor Y-Ba-Cu-0. *Physical Review B*, 35(16):8846–8849, 1987.

[102] M. F. Crommie, L. C. Bourne, A. Zettl, M. L. Cohen, and A. Stacy. Tunneling measurement of the energy gap in Y-Ba-Cu-0. *Physical Review B*, 35(16):8853–8855, 1987.

[103] M. Fogelström, D. Rainer, and J. A. Sauls. Tunneling into Current-Carrying Surface States of High-T_c Superconductors. *Physical Review Letters*, 79(2):281–284, 1997.

[104] G. Deutscher. Andreev-Saint-James reflections: A probe of cuprate superconductors. *Reviews of modern physics*, 77(1):109–135, 2005.

[105] N. Perrin and C. Vanneste. Response of superconducting films to a periodic optical irradiation. *Physical Review B*, 28(9):5150–5159, 1983.

[106] M. Lindgren, M. Currie, C. A. Williams, T. Y. Hsiang, P. M. Fauchet, R. Sobolewski, S. H. Moffat, R. A. Hughes, J. S. Preston, and F. A. Hegmann. Intrinsic picosecond response times of Y-Ba-Cu-O superconducting photodetectors. *Applied Physics Letters*, 74(6):853–855, 1999.

[107] C. G. Levey, S. Etemad, and A. Inam. Optically detected transient thermal response of high T_c epitaxial films. *Applied Physics Letters*, 60(1):126–128, 1992.

[108] M. Lindgren, H. Ahlberg, A. Larsson, S. T. Eng, and M. Danerud. Ultrafast IR detector response in high T_c superconducting thin films. *Physica Scripta*, 44(1):105–107, 1991.

[109] A. Frenkel, M. A. Saifi, T. Venkatesan, P. England, X. D. Wu, and A. Inam. Optical response of nongranular high-T_c $Y_1Ba_2Cu_3O_{7-x}$ superconducting thin films. *Journal of Applied Physics*, 67(6):3054–3068, 1990.

[110] A. Frenkel, M. A. Saifi, T. Venkatesan, C. Lin, X. D. Wu, and A. Inam. Observation of fast nonbolometric optical response of nongranular high T_c $Y_1Ba_2Cu_3O_{7-x}$ superconducting thin films. *Applied Physics Letters*, 54(16):1594–1596, 1989.

[111] A. Frenkel. High Temperature Superconducting Thin Films as Broadband Optical Detectors. *Physica C*, 180:251–258, 1991.

[112] T. T. M. Palstra, B. Batlogg, R. B. van Dover, L. F. Schneemeyer, and J. V. Waszczak. Critical currents and thermally activated flux motion in hightemperature superconductors. *Applied Physics Letters*, 54(8):763–765, 1989.

[113] A. Frenkel, E. Clausen, C. C. Chang, T. Venkatesan, P. S. D. Lin, X. D. Wu, A. Inam, and B. Lalevic. Imaging of current-induced superconducting-resistive transitions by scanning electron microscopy in laser-deposited superconducting thin films of $Y_1Ba_2Cu_3O_{7-x}$. *Applied Physics Letters*, 55(9):911–913, 1989.

[114] Y. Yeshurun and A. P. Malozemoff. Giant Flux Creep and Irreversibility in an Y-Ba-Cu-O Crystal: An Alternative to the Superconducting-Glass Model. *Physical Review Letters*, 60(21):2202–2205, 1988.

[115] K. Enpuku, T. Kisu, R. Sako, K. Yoshida, M. Takeo, and K. Yamafuji. Effect of flux creep on current-voltage characteristics of superconducting Y-Ba-Cu-O thin films. *Japanese Journal of Applied Physics*, 28(6):L991–L993, 1989.

[116] A. M. Kadin, M. Leung, A. D. Smith, and J. M. Murduck. Photofluxonic detection: A new mechanism for infrared detection in superconducting thin films. *Applied Physics Letters*, 57(26):2847–2849, 1990.

[117] A. Rothwarf and B. N. Taylor. Measurement of recombination lifetimes in superconductors. *Physical Review Letters*, 19(1):27–30, 1967.

[118] E. N. Grossman, D. G. McDonald, and J. E. Sauvageau. Measurement of recombination lifetimes in superconductors. *IEEE Transactions on Magnetics*, 27(2):2677–2680, 1991.

[119] N. Bluzer. Analysis of quantum superconducting kinetic inductance photodetectors. *Journal of applied physics*, 78(12):7340–7351, 1995.

[120] R. S. Nebosis, M. A. Heusinger, W. Schatz, K. F. Renk, G. N. Gol'tsman, B. S. Karasik, A. D. Semenov, and G. M. Gershenzon. Ultrafast photoresponse of a structured $YBa_2Cu_3O_{7-\delta}$ thin film to ultrashort FIR laser pulses. *IEEE Transactions on applied superconductivity*, 3(1):2160–2162, 1993.

[121] K. S. Il'in and M. Siegel. Microwave mixing in microbridges made from $YBa_2Cu_3O_{7-d}$ thin films. *Journal of Applied Physics*, 92(1):361–369, 2002.

[122] T. Venkatesan, X. D. Wu, B. Dutta, A. Inam, M. S. Hegde, D. M. Hwang, C. C. Chang, L. Nazar, and B. Wilkens. High-temperature superconductivity in ultrathin films of $Y_1Ba_2Cu_3O_{7-x}$. *Applied Physics Letters*, 54(6):581–583, 1989.

[123] B. Holzapfel, B. Roas, L. Schultz, P. Bauer, and G. Saemann-Ischenko. Off-axis laser deposition of $YBa_2Cu_3O_{7-x}$ thin films. *Applied Physics Letters*, 61(26):3178–3180, 1992.

[124] T. Haugan, P. N. Barnes, L. Brunke, I. Maartense, and J. Murphy. Effect of O_2 partial pressure on $YBa_2Cu_3O_{7-x}$ thin film growth by pulsed laser deposition. *Physica C*, 397:47–57, 2003.

[125] V. Ban and D. Kramer. Thin films of semiconductors and dielectrics produced by laser evaporation. *Journal of Materials Science*, 5(11):978–982, 1970.

[126] J. Brannon. *Excimer Laser, Ablation and etching*. The American Vacuum Society Education Committee, 335 East 45th Street, New York, New York 10017, 1993.

[127] P. Probst. *High-T_c thin films for THz detectors*. Diplomarbeit am Institut für Mikro- und Nanoelektronische Systeme, 2009.

[128] M. Winter. *Deposition durch Laserablation und Sauerstoffbeladung von Mehrfachschichten aus oxidischen Supraleitern und Isolatoren*, 1997. Diplomarbeit, Institut für Elektrotechnische Grundlagen der Informatik, Universität Karlsruhe.

[129] Lambda Physik GmbH Göttingen. *Lambda Physik Excimer Lasers*, 1993.

[130] R. Wördenweber. Growth of high-T_c thin films. *Superconductor Science and Technology*, 12:R86–R102, 1999.

[131] I.-S. Kim, H.-R. Lim, and Y. K. Park. Epitaxial Growth of YBCO Thin Films on Al_2O_3 Substrates by Pulsed Laser Deposition. *IEEE Transactions on Applied Superconductivity*, 9(2):1649–1652, 1999.

[132] S. Y. Lee. Characterization of the interface between laser ablated YBCO film and buffered sapphire substrate. *Journal of Alloys and Compounds*, 251:41–43, 1997.

[133] F. Wang and R. Wördenweber. Large-area epitaxial CeO_2 buffer layers on sapphire substrates for the growth of high quality $YBa_2Cu_3O_7$ films. *Thin solid films*, 227:200–204, 1993.

[134] D. H. Lowndes, D. P. Norton, and J. D. Budai. Superconductivity in nonsymmetric epitaxial $YBa_2Cu_3O_{7-x}/PrBa_2Cu_3O_{7-x}$ superlattices: the superconducting behaviour of Cu-O bilayers. *Physical Review Letters*, 65(9):1160–1163, 1990.

[135] F. A. Hegmann and J. S. Preston. Origin of the fast photoresponse of epitaxial $YBa_2Cu_3O_{7-\delta}$ thin films. *Physical Review B*, 48(21):16023–16039, 1993.

[136] J. Ye and K. Nakamura. Quantitative structure analysis of $YBa_2Cu_3O_{7-\delta}$ thin films: Determination of oxygen content from x-ray-diffraction patterns. *Physical Review B*, 48(10):7554–7564, 1993.

[137] D. R. Lide. *Handbook of Chemistry and Physics, 75th Edition.* New York: CRC Press, 1996-1997.

[138] M. Tinkham. *Introduction to Superconductivity second edn.* McGraw-Hill, Inc., New York, 1996.

[139] K.-H. Bennemann and J. B. Ketterson. *Superconductivity: Volume 1: Conventional and Unconventional Superconductors.* Springer; 1 edition, 2008.

[140] L. G. Aslamazov and A. I. Larkin. The influence of fluctuation pairing of electrons on the conductivity of normal metal. *Physical Letters A*, 26(6):238–239, 1968.

[141] B. Oh, K. Char, A. D. Kent, M. Naito, M. R. Beasly, T. H. Geballe, R. H. Hammond, A. Kapitulnik, and J. M. Graybeal. Upper critical field, fluctuation conductivity, and dimensionality of $YBa_2Cu_3O_{7-x}$. *Physical Review B*, 37(13):7861–7864, 1988.

[142] V. L. Ginzburg and L. D. Landau. On the theory of superconductivity. *Zh. Eksper. Teor. Fiz*, 20:1064–1082, 1950.

[143] M. M. Fang, V. G. Kogan, D. K. Finnemore, J. R. Clem, L. S. Chumbley, and D. E. Farrell. Possible twin-boundary effect upon the properties of high-T_c superconductors. *Physical Review B*, 37(4):2334–2337, 1988.

[144] M. Tinkham. Resistive Transition of High-Temperature Superconductors. *Physical Review Letters*, 61(14):1658–1661, 1988.

[145] E. Helfand and N. R. Werthamer. Temperature and Purity Dependence of the Superconducting Critical Field H_{c2} II. *Physical Review*, 147(1):288–294, 1966.

[146] H. C. Yang, L. M. Wang, and H. E. Horng. Characteristics of flux pinning in $YBa_2Cu_3O_y$ /$PrBa_2Cu_3O_y$ superlattices. *Physical Review B*, 59(13):8956–8961, 1999.

[147] N. R. Werthamer, E. Helfand, and P. C. Hohenberg. Temperature and Purity Dependence of the Superconducting Critical Field H_{c2} III. Electron spin and spin orbit effects. *Physical Review*, 147(1):295–302, 1966.

[148] A. Semenov, B. Günther, U. Böttger, H.-W. Hübers, H. Bartolf, A. Engel, A. Schilling, K. Il'in, M. Siegel, R. Schneider, D. Gerthsen, and N. A. Gippius. Optical and transport properties of ultrathin NbN films and nanostructures. *Physical Review B*, 80(054510), 2009.

[149] A. Scheuring. *Ultrabreitbandige Strahlungseinkopplung in THz-Detektoren*, 2013. PhD thesis, Institut für Mikro- und Nanoelektronische Systeme, Karlsruher Institut für Technologie (KIT).

[150] CST Microwave Studio. Release Version 2011.05. http://www.cst.com, 2011.

[151] R. Müller, A. Hoehl, A. Serdyukov, G. Ulm, J. Feikes, M. Ries, and G. Wüstefeld. The Metrology Light Source of PTB - a Source for THz Radiation. *Journal of Infrared, Millimeter and Terahertz waves*, 32(6):742–753, 2011.

[152] A.-S. Müller, I. Birkel, S. Casalbuoni, B. Gasharova, E. Huttel, Y.-L. Mathis, D. A. Moss, N. J. Smale, P. Wesolowski, E. Bruendermann, T. Bueckle, and M. Klein. Characterizing THz coherent synchrotron radiation at the ANKA storage ring. *Proceedings of EPAC08*, WEPC046:2091–2093, 2008.

[153] A. D. Semenov, H. Richter, H.-W. Hübers, B. Günther, A. Smirnov, K. S. Il'in, M. Siegel, and J. P. Karamarkovic. Terahertz Performance of Integrated Lens Antennas With a Hot-Electron Bolometer. *IEEE Transactions on microwave theory and techniques*, 55(2):239–247, 2007.

[154] A. Engel, A. Schilling, K. Il'in, and M. Siegel. Dependence of count rate on magnetic field in superconducting thin-film TaN single-photon detectors. *Physical Review B*, 86(140506):140506–1–140506–4, 2012.

[155] B. Pümpin, H. Keller, W. Kündig, W. Odermatt, I. M. Savic, J. W. Schneider, H. Simmler, P. Zimmermann, J. G. Bednorz, Y. Maeno, K. A. Müller, C. Rossel, E. Kaldis, S. Rusiecki, W. Assmus, and J. Kowalewski. Measurements of the London penetration depths in $YBa_2Cu_3O_x$ by means of muon spin rotation (μSR) experiments. *Physica C*, 162-164:151–152, 1989.

[156] TOPTICA Photonics. http://www.toptica.com.

[157] S. Wünsch, T. Ortlepp, E. Crocoll, F. H. Uhlmann, and M. Siegel. Cryogenic Semiconductor Amplifier for RSFQ-Circuits With High Data Rates at 4.2 K. *IEEE Transactions on Applied Superconductivity*, 19(3):574–579, 2009.

[158] A. V. Sergeev, A. D. Semenov, P. Kouminov, V. Trifonov, I. G. Goghidze, B. S. Karasik, G. N. Gol'tsman, and E. M. Gershenzon. Transparency of a $YBa_2Cu_3O_7$-film/ substrate interface for thermal phonons measured by means of voltage response to radiation. *Physical Review B*, 49(13):9091–9096, 1994.

[159] Y. P. Gousev, A. D. Semenov, R. S. Nebosis, E. V. Pechen, A. V. Varlashkin, and K. F. Renk. Broad-band coupling of THz radiation to an $YBa_2Cu_3O_{7-x}$ hot-electron bolometer mixer. *Superconductor Science and Technology*, 9:779–787, 1996.

[160] A. D. Semenov, M. A. Heusinger, J. H. Hoffmann, and K. F. Renk. Autocorrelation Study of Quasiparticle Relaxation in a $YBa_2Cu_3O_{7-\delta}$ Film. *Journal of Low Temperature Physics*, 105:305–310, 1996.

[161] R. C. Jones. The General Theory of Bolometer Performance. *Journal of the Optical Society*, 43(1):1–14, 1953.

[162] P. F. Goldsmith. *Quasioptical Systems*. Institute of Electrical and Electronics Engineers, Inc., New York, 1998.

[163] S. G. Doettinger, R. P. Huebener, R. Gerdemann, A. Kühle, S. Anders, T. G. Träuble, and J. C. Villégier. Electronic Instability at High Flux-Flow Velocities in High-T_c Superconducting Films. *Physical Review Letters*, 73(12):1691–1694, 1994.

[164] A. I. Larkin and Yu. N. Ovchinnikov. Nonlinear conductivity of superconductors in the mixed state. *Sov. Phys. JETP*, 41(5):960–965, 1975.

[165] D. Y. Vodolazov and F. M. Peeters. Rearrangement of the vortex lattice due to instabilities of vortex flow. *Physical Review B*, 76(014521):014521-1–014521-9, 2007.

[166] G. R. Berdiyorov, M. V. Milosevic, and F. M. Peeters. Kinematic vortex-antivortex lines in strongly driven superconducting stripes. *Physical Review B*, 79(184506):184506-1–184506-8, 2009.

[167] D. F. Filipovic, S. S. Gearhart, and G. M. Rebeiz. Double-slot antennas on extended hemispherical and elliptical silicon dielectric lenses. *IEEE Transactions on Microwave Theory and Techniques*, 41(10):1738–1749, 1993.

[168] SHF Communication Technologies AG. Tutorial Note 2 Microwave Connectors. http://www.shf.de, 2009.

[169] Y.-L. Mathis, B. Gasharova, and D. Moss. Terahertz Radiation at ANKA, the new synchrotron light source in Karlsruhe. *Journal of Biological Physics*, 29:313–318, 2003.

[170] R. Klein, G. Brandt, R. Fliegauf, A. Hoehl, R. Müller, R. Thornagel, and G. Ulm. Operation of the Metrology Light Source as a primary radiation source standard. *Physical Review Special Topics - Accelerators and beams*, 11(110701):1–10, 2008.

[171] F. Ballout, J.-S. Samson, D. A. Schmidt, E. Bründermann, Y.-L. Mathis, B. Gasharova, A. D. Wieck, and M. Havenith. Non-invasive nano-imaging of ion implanted and activated copper in silicon. *Journal of Applied Physics*, 110(024307):024307-1–024307-7, 2011.

[172] G. L. Carr, M. C. Martin, W. R. McKinney, K. Jordon, G. R. Neil, and G. P. Williams. High-power terahertz radiation from relativistic electrons. *Nature*, 420:153–156, 2002.

[173] T. Nakazato, M. Oyamada, N. Niimura, S. Urasawa, O. Konno, A. Kagaya, R. Kato, T. Kamiyama, Y. Torizuka, T. Nanba, Y. Kondo, Y. Shibata, K. Ishi, T. Ohsaka, and M. Ikezawa. Observation of coherent synchrotron radiation. *Physical Review Letters*, 63(12):1245–1248, 1989.

[174] R. W. Schoenlein, S. Chattopadhyay, H. H. W. Chong, T. E. Glover, P. A. Heimann, C. V. Shank, A. A. Zholents, and M. S. Zolotorev. Generation of Femtosecond Pulses of Synchrotron Radiation. *Science*, 287(5461):2237–2240, 2000.

[175] M. Shimada, M. Katoh, S. Kimura, A. Mochihashi, M. Hosaka, Y. Takashima, T. Hara, and T. Takahashi. Intense Terahertz Synchrotron Radiation by Laser Bunch Slicing at UVSOR-II Electron Storage Ring. *Japanese Journal of Applied Physics*, 46(12):7939–7944, 2007.

[176] K. Holldack, S. Khan, R. Mitzner, and T. Quast. Femtosecond Terahertz Radiation from Femtoslicing at BESSY. *Physical Review Letters*, 96:054801, 2006.

[177] S. Bielawski, C. Evain, T. Hara, M. Hosaka, M. Katoh, S. Kimura, A. Mochihashi, M. Shimada, C. Szwaj, T. Takahashi, and Y. Takashima. Tunable narrowband terahertz emission from mastered laser-electron beam interaction. *Nature physics*, 4:390–393, 2008.

[178] M. Shimada, M. Katoh, M. Adachi, T. Tanikawa, S. Kimura, M. Hosaka, N. Yamamoto, Y. Takashima, and T. Takahashi. Transverse-longitudinal coupling effect in laser bunch slicing. *Physical Review Letters*, 103:144802, 2009.

[179] C. Evain, C. Szwaj, S. Bielawski, M. Hosaka, Y. Takashima, M. Shimada, S. Kimura, M. Katoh, A. Mochihashi, T. Takahashi, and T. Hara. Laser-induced narrowband coherent synchrotron radiation: Efficiency versus frequency and laser power. *Physical Review Special Topics: Accelerators and beams*, 13:090703, 2010.

[180] N. Hiller, A. Hofmann, E. Huttel, V. Judin, B. Kehrer, M. Klein, M. Marsching, and A.-S. Müller. Status of Bunch Deformation and Lengthening Studies at the ANKA Storage Ring. *Proceedings of IPAC2011, San Sebastián, Spain*, THPC021:2951–2953, 2011.

[181] Hamatsu Photonics Deutschland GmbH. http://sales.hamamatsu.com/en/products/system-division/ultra-fast/streak-systems.php.

[182] F. Sannibale, J. M. Byrd, Á. Loftsdóttir, M. Venturini, M. Abo-Bakr, J. Feikes, K. Holldack, P. Kuske, G. Wüstefeld, H.-W. Hübers, and R. Warnock. A Model Describing Stable Coherent Synchrotron Radiation in Storage Rings. *Physical Review Letters*, 93(9):094801, 2004.

[183] J. Barros, L. Manceron, J.-B. Brubach, G. Creff, C. Evain, M.-E. Couprie, A. Loulergue, L. Nadolski, M.-A. Tordeux, and P. Roy. Toward highly stable Terahertz Coherent Synchrotron Radiation at the synchrotron SOLEIL. *Journal of Physics: Conference Series*, 359(012002):1–5, 2012.

[184] K. Holldack, T. Kachel, S. Khan, R. Mitzner, and T. Quast. Characterization of laser-electron interaction at the BESSY II femtoslicing source. *Physical Review Special Topics: Accelerators and Beams*, 8:040704, 2005.

[185] S. Khan, K. Holldack, T. Kachel, R. Mitzner, and T. Quast. Femtosecond undulator radiation from sliced electron bunches. *Physical Review Letters*, 97:074801, 2006.

[186] E. Roussel, C. Evain, M. Le Parquier, S. Bielawski, C. Szwaj, M. Adachi, H. Zen, T. Tanikawa, M. Katoh, M. Hosaka, and N. Yamamoto. Laser-induced CSR: toward a probe to explore wakefields in storage rings? *International Particle Accelerator Conference 2012*, 2012.

[187] G.-S. Park, Y. H. Kim, H. Han, J. K. Han, and J. Ahn. *Convergence of Terahertz Sciences in Biomedical Systems*. Springer; Auflage: 2013, 2012.

[188] F. J. Duarte (Ed.). *Tunable Lasers Handbook, Chapter 9*. Academic, New York, 1995.

[189] J. Feldhaus, J. Arthur, and J. B. Hastings. X-ray free-electron lasers. *Journal of Physics B: Atomic, Molecular and Optical Physics*, 38:S799–S819, 2005.

[190] Isoyama Group. Institute of Scientific and Industrial Research. http://www.phys.sci.osaka-u.ac.jp, 2010.

[191] G. Scalari, C. Walther, M. Fischer, R. Terazzi, H. Beere, D. Ritchie, and J. Faist. THz and sub-THz quantum cascade lasers. *Laser and Photonics Review*, 3:45–66, 2009.

[192] S. Kumar. Recent Progress in Terahertz Quantum Cascade Lasers. *IEEE Journal of Selected Topics in Quantum Electronics*, 17:38–47, 2011.

[193] B. S. Williams, S. Kumar, Q. Hu, and J. L. Reno. High-power terahertz quantum-cascade lasers. *Electronics Letters*, 42:89–90, 2006.

[194] A. Barkan, F. K. Tittel, D. M. Mittleman, R. Dengler, P. H. Siegel, G. Scalari, L. Ajili, J. Faist, H. E. Beere, E. H. Linfield, A. G. Davies, and D. A. Ritchie. Linewidth and tuning characteristics of terahertz quantum cascade lasers. *Optics Letters*, 29:575–577, 2004.

[195] M. Tonouchi. Cutting-edge terahertz technology. *Nature Photonics*, 1:97–105, 2007.

[196] G. Scalari, L. Ajili, J. Faist, H. Beere, E. H. Linfield, D. Ritchie, and G. Davies. Far-infrared (λ=87 μm) bound-to-continuum quantumcascade lasers operating up to 90 K. *Applied Physics Letters*, 82(19):3165–3167, 2003.

Own Publications

[PIE$^+$12] P. Probst, K. Il'in, A. Engel, A. Semenov, H.-W. Hübers, J. Hänisch, B. Holzapfel, and M. Siegel. Magnetoresistivity of thin YBa$_2$Cu$_3$O$_{7-d}$ films on sapphire substrate. *Physica C: Superconductivity*, 479:173–175, 2012.

[PSH$^+$11] P. Probst, A. Scheuring, M. Hofherr, D. Rall, S. Wünsch, K. Il'in, M. Siegel, A. Semenov, A. Pohl, H.-W. Hübers, V. Judin, A.-S. Müller, A. Hoehl, R. Müller, and G. Ulm. YBa$_2$Cu$_3$O$_{7-d}$ quasioptical detectors for fast time-domain analysis of terahertz synchrotron radiation. *Applied Physics Letters*, 98(043504), 2011.

[PSH$^+$12a] A. Pohl, A. Semenov, A. Hoehl, R. Müller, G. Ulm, J. Feikes, M. Ries, G. Wüstefeld, P. Probst, A. Scheuring, M. Hofherr, S. Wünsch, K. Il'in, M. Siegel, and H.-W. Hübers. Measurements of the time jitter of coherent terahertz synchrotron radiation with a superconducting detector. *IEEE Transactions on microwave theory and techniques*, 1:1–2, 2012.

[PSH$^+$12b] P. Probst, A. Scheuring, M. Hofherr, S. Wünsch, K. Il'in, A. Semenov, H.-W. Hübers, V. Judin, A.-S. Müller, J. Hänisch, B. Holzapfel, and M. Siegel. Superconducting YBa$_2$Cu$_3$O$_{7-d}$ Thin Film Detectors for Picosecond THz Pulses. *Journal of Low Temperature Physics*, 167(5):898–903, 2012.

[PSR$^+$12] P. Probst, A. Semenov, M. Ries, A. Hoehl, P. Rieger, A. Scheuring, V. Judin, S. Wünsch, K. Il'in, N. Smale, Y.-L. Mathis, R. Müller, G. Ulm, G. Wüstefeld, H.-W. Hübers, J. Hänisch, B. Holzapfel, M. Siegel, and A.-S. Müller. Non-thermal response of YBa$_2$Cu$_3$O$_{7-d}$ thin films to picosecond THz pulses. *Physical Review B*, 85(174511), 2012.

[RPH$^+$10] D. Rall, P. Probst, M. Hofherr, S. Wünsch, K. Il'in, U. Lemmer, and M. Siegel. Energy relaxation time in NbN and YBCO thin films under optical irradiation. *Journal of Physics: Conference Series*, 234(042029), 2010.

[SDV$^+$13] A. Scheuring, P. Dean, A. Valavanis, A. Stockhausen, P. Thoma, M. Salih, S. P. Khanna, S. Chowdhury, J. D. Cooper, A. Grier, S. Wünsch, K. Il'in, E. H. Linfield, A. G. Davies, and M. Siegel. Transient analysis of THz-QCL pulses using NbN and YBCO superconducting detectors. *IEEE Transactions on Terahertz Science and Technology*, 3(2):172–179, 2013.

"""

[STD+13] A. Scheuring, P. Thoma, J. Day, K. Il'in, J. Hänisch, B. Holzapfel, and M. Siegel. Thin Pr-Ba-Cu-O Film Antenna-Coupled THz Bolometers for Room Temperature Operation. *IEEE Transactions on Terahertz Science and Technology*, 3(1):103–109, 2013.

[TRS+13] P. Thoma, J. Raasch, A. Scheuring, M. Hofherr, K. Il'in, S. Wünsch, A. Semenov, H.-W. Hübers, V. Judin, A.-S. Müller, N. Smale, J. Hänisch, B. Holzapfel, and M. Siegel. Highly responsive Y-Ba-Cu-O thin film THz detectors with picosecond time resolution. *IEEE Transactions on Applied Superconductivity*, 23(3):2400206, 2013.

[TSH+12] P. Thoma, A. Scheuring, M. Hofherr, S. Wünsch, K. Il'in, N. Smale, V. Judin, N. Hiller, A.-S. Müller, A. Semenov, H.-W. Hübers, and M. Siegel. Real-time measurements of picosecond THz pulses by an ultra-fast $YBa_2Cu_3O_{7-d}$ detection system. *Applied Physics Letters*, 101(142601), 2012.

[TSW+13] P. Thoma, A. Scheuring, S. Wünsch, K. Il'in, A. Semenov, H.-W. Hübers, V. Judin, A.-S. Müller, N. Smale, M. Adachi, S. Tanaka, S. Kimura, M. Katoh, N. Yamamoto, M. Hosaka, E. Roussel, C. Szwaj, S. Bielawski, and M. Siegel. High-speed Y-Ba-Cu-O direct detection system for monitoring picosecond THz pulses. *IEEE Transactions on Terahertz Science and Technology*, 3(1):81–86, 2013.

International Conferences

[a-ASC] P. Thoma, J. Raasch, A. Scheuring, M. Hofherr, K. Il'in, S. Wünsch, A. Semenov, H.-W. Hübers, V. Judin, A.-S. Müller, N. Smale, J. Hänisch, B. Holzapfel, and M. Siegel. Thin YBCO film THz detector with picosecond time resolution and large dynamic range. *Invited talk at the Applied Superconductivity Conference 2012 (ASC)*, Portland, Oregon, US, 7-12 of October 2012.

[b-ISSTT] P. Thoma, A. Scheuring, M. Hofherr, S. Wünsch, K. Il'in, A. Pohl, A. Semenov, H.-W. Hübers, V. Judin, A.-S. Müller, A. Hoehl, R. Müller, G. Ulm, J. Hänisch, B. Holzapfel, and M. Siegel. YBa_2Cu_3O7-d high-speed detectors for picosecond THz pulses. *23rd International Symposium on Space Terahertz Technology (ISSTT)*, Tokyo, Japan, 2-4 of April 2012.

[c-KRYO] A. Scheuring, A. Stockhausen andP. Probst, A. Schmid, S. Wünsch, K. Il'in, and M. Siegel. Design, simulation and characterization of single-pixel and multi-pixel bolometric THz detectors. *Meeting on Cryoelectronic devices (Kryo 2011)*, Grenoble, France, 2-4 of October 2011.

[d-EUCAS] A. Pohl, P. Probst, A. Scheuring, K. Il'in, A. Semenov, H.-W. Hübers, M. Siegel, A. Hoehl, R. Müller, G. Wüstefeld, and G. Ulm. Fast superconducting detector for analysing jitter of THz synchrotron radiation pulses. *European Conference on Applied Superconductivity (EUCAS)*, The Hague, Netherlands, 18-23 of September 2011.

[e-VORTEX] P. Probst, A. Scheuring, M. Hofherr, S. Wünsch, K. Il'in, A. Semenov, H.-W. Hübers, V. Judin, A.-S. Müller, J. Hänisch, B. Holzapfel, and M. Siegel. Magnetoresistivity of thin YBa_2Cu_3O7-d films on sapphire. *Vortex Matter in Nanostructured Superconductors (VORTEX VII)*, Rhodos, Greece, 10-16 of September 2011.

[f-WIRMS] P. Probst, A. Scheuring, M. Hofherr, S. Wünsch, K. Il'in, A. Pohl, A. Semenov, H.-W. Hübers, V. Judin, A.-S. Müller, J. Hänisch, B. Holzapfel, A. Hoehl, R. Müller, G. Ulm, and M. Siegel. Synchrotron pulse detection with a YBa_2Cu_3O7-d thin-film THz detector. *6th Workshop on Infrared Spectroscopy and Microscopy with Accelerator-Based Sources (WIRMS)*, Trieste, Italy, 4-7 of September 2011.

[g-LTD] P. Probst, A. Scheuring, M. Hofherr, S. Wünsch, K. Il'in, A. Semenov, H.-W. Hübers, V. Judin, A.-S. Müller, J. Hänisch, B. Holzapfel, and M. Siegel. THz pulse detection with picosecond time resolution by a $YBa_2Cu_3O7 - d$ thin-film detector. *14th International Workshop on Low Temperature Detectors (LTD)*, Heidelberg, Germany, 1-5 of August 2011.

[h-KRYO] P. Probst, A. Scheuring, M. Hofherr, A. Stockhausen, D. Rall, S. Wünsch, K. Il'in, A. Semenov, H.-W. Hübers, V. Judin, A.-S. Müller, J. Hänisch, B. Holzapfel, and M. Siegel. High-speed YBCO detectors for resolving picosecond THz pulses. *Meeting on Cryoelectronic devices (Kryo 2010)*, Zeuthen, Germany, 3-5 of October 2010.

[i-THzsymp] P. Probst, D. Rall, A. Scheuring, M. Hofherr, S. Wünsch, K. Il'in, A. Semenov, H.-W. Hübers, V. Judin, A.-S. Müller, and M. Siegel. Superconducting YBCO hot-electron bolometer detectors for picosecond time resolution of THz pulses. *Best poster award at 456. Wilhelm und Else Heraeus Seminar, THz radiation: Generation, Detection and Applications*, Bad Honnef, Germany, 18-21 of April 2010.

[j-DPG] P. Probst, D. Rall, M. Hofherr, S. Wünsch, K. Il'in, and M. Siegel. Energy relaxation processes in YBCO thin films studied by frequency and time-domain techniques. *DPG Frühjahrstagung der Sektion Kondensierte Materie (SKM)*, Regensburg, Germany, 21-26 of March 2010.

Supervised Student Theses

[Day11] Day, Julia. *Supraleitende Eigenschaften in Sub-Mikrometer-Strukturen aus YBCO*, 2011. Studienarbeit, Institut für Mikro- und Nanoelektronische Systeme, Karlsruher Institut für Technologie (KIT).

[Day12] Day, Julia. *Optimierung von PBCO THz Detektoren*, 2012. Diploma thesis, Institut für Mikro- und Nanoelektronische Systeme, Karlsruher Institut für Technologie (KIT).

[de 11] de Zabala Wissing, Mikel. *Entwicklung von THz-Raumtemperaturdetektoren basierend auf PBCO*, 2011. Studienarbeit, Institut für Mikro- und Nanoelektronische Systeme, Karlsruher Institut für Technologie (KIT).

[Kal12] Kalteisen, Diana. *Optimierung der supraleitenden Eigenschaften von YBCO Dünnschichten*, 2012. Bachelor thesis, Institut für Mikro- und Nanoelektronische Systeme, Karlsruher Institut für Technologie (KIT).

[Raa12] Raasch, Juliane. *Detektionsmechanismus in sub-Mikrometer YBCO Strukturen*, 2012. Diploma thesis, Institut für Mikro- und Nanoelektronische Systeme, Karlsruher Institut für Technologie (KIT).

[Wer10] Wernerus, Holger. *Untersuchung laserablatierter Goldschichten für THz-Detektoren*, 2010. Bachelor thesis, Institut für Mikro- und Nanoelektronische Systeme, Karlsruher Institut für Technologie (KIT).

Karlsruher Schriftenreihe zur Supraleitung
(ISSN 1869-1765)

Herausgeber: Prof. Dr.-Ing. M. Noe, Prof. Dr. rer. nat. M. Siegel

**Die Bände sind unter www.ksp.kit.edu als PDF frei verfügbar oder
als Druckausgabe bestellbar.**

Band 001
Christian Schacherer
**Theoretische und experimentelle Untersuchungen zur
Entwicklung supraleitender resistiver Strombegrenzer.** 2009
ISBN 978-3-86644-412-6

Band 002
Alexander Winkler
Transient behaviour of ITER poloidal field coils. 2011
ISBN 978-3-86644-595-6

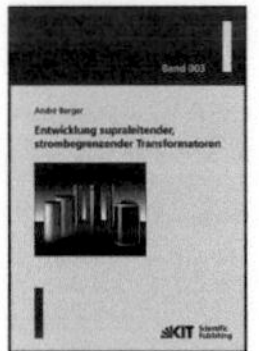

Band 003
André Berger
**Entwicklung supraleitender, strombegrenzender
Transformatoren.** 2011
ISBN 978-3-86644-637-3

Band 004
Christoph Kaiser
**High quality Nb/Al-AlOx/Nb Josephson junctions.
Technological development and macroscopic quantum
experiments.** 2011
ISBN 978-3-86644-651-9

Band 005
Gerd Hammer
**Untersuchung der Eigenschaften von planaren Mikrowellen-
resonatoren für Kinetic-Inductance Detektoren bei 4,2 K.** 2011
ISBN 978-3-86644-715-8

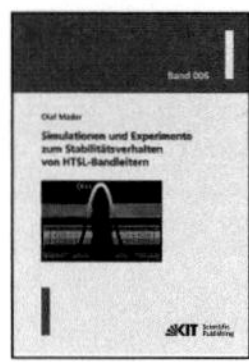

Band 006
Olaf Mäder
**Simulationen und Experimente zum Stabilitätsverhalten
von HTSL-Bandleitern.** 2012
ISBN 978-3-86644-868-1

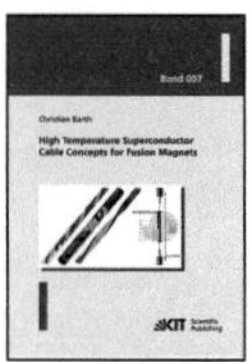

Band 007
Christian Barth
**High Temperature Superconductor Cable Concepts for
Fusion Magnets.** 2013
ISBN 978-3-7315-0065-0

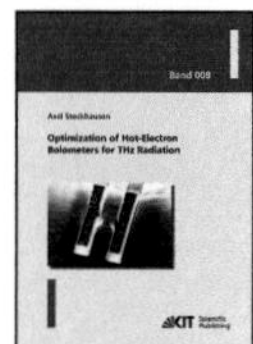

Band 008
Axel Stockhausen
Optimization of Hot-Electron Bolometers for THz Radiation. 2013
ISBN 978-3-7315-0066-7

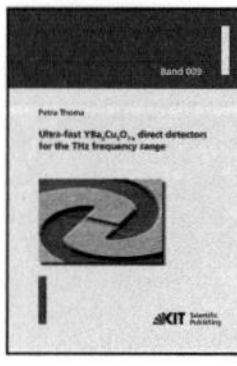

Band 009
Petra Thoma
**Ultra-fast YBa$_2$Cu$_3$O$_{7-x}$ direct detectors for the THz frequency
range.** 2013
ISBN 978-3-7315-0070-4